Soviet Space Secrets

Hidden stories
of the Space Race

by

Dominic Phelan

CONTENTS

Introduction

The late 1980s were an exciting time for space enthusiasts as they witnessed a re-emergence of the 'space race' between the United States and the Soviet Union. In the two-and-a-half years that NASA was grounded by the *Challenger* accident the Soviets dominated headlines with the *Mir* space station, robotic probes to Halley's Comet and their own shuttle *Buran*.

In our era of information overload, it seems bizarre to realise there was once a time when basic facts about Soviet spaceflight were hard to find. That Moscow could keep major rocket accidents and cosmonaut deaths a secret is even more unbelievable.

My first glimpse behind this mysterious 'space curtain' was a 1986 cover story in *National Geographic* magazine. This was a revelation because it revealed the existence of the underground pursuit of 'space sleuthing'. These amateur detectives - some might call them part-time spies! - had a passion for hunting down and publishing the hidden secrets of Soviet spaceflight.

This opening up of new chapters in the history of the Space Race coincided with Mikhail Gorbachev's new policy of *glasnost* (openness), which was finally confirming the country's darkest secrets to those behind the Iron Curtain. As the USSR itself collapsed, suppressing embarrassing facts no longer seemed to matter to a superpower in terminal decline.

I made my own pilgrimage to the Yuri Gagarin Cosmonaut Training Centre in 2003 as part of an officially organised press tour. In an example of idiosyncratic Russian planning we arrived on the same day its new director was being sworn in at a separate closed ceremony and the facility was a virtual ghost town.

In a scene unthinkable only a decade before, we were left to roam the centre without supervision. If we grew bored listening to our tour guide, we simply wandered-off to explore its run-down hangars and dusty corridors alone.

After this eye-opening look around I returned home confident enough to write about the subject with authority. These articles, published together here for the first time, show the real story was often more interesting than the propaganda myth.

Dominic Phelan

The Purge of Alexander Scherschevsky

Alexander Boris Scherschevsky was born into a wealthy St. Petersburg Jewish family on 22nd October 1894 and was privately educated before entering the city's Polytechnic Institute in 1913 to study Mechanical Engineering. Although he had developed an interest in aeronautics as a young boy (he even organised a model-aircraft club at school), any hopes of becoming a flyer himself were dashed when he had to leave a pilot training course at the local Aero Club because of poor eyesight.

He found a job at the Lebedev aircraft factory but then decided to move to Germany in 1919 to study at the Berlin Technical University. His lecturers there are reported to have included Albert Einstein and Max Plank [1]. Although the young Russian was sympathetic to the Communist regime back home he decided to stay in Berlin when his course ended. As this broke the terms of his Soviet state-sponsored 'Kommandirovka' visa he then effectively became what his contemporary Willy Ley described as a 'Bolshevist accidently in disgrace' [2].

Although he worked at the patent office of Berlin's Rohrbach aircraft factory for a few years, his main source of income during the 1920s was from writing freelance aviation articles for the German science and aviation press. During this period he was also commissioned to compile the Russian sections for a seven volume German aviation dictionary. Luckily, he had been reading the works of Russian spaceflight pioneer Konstantin Tsiolkovsky since 1911 and was the first person to mention him to a German audience during a public lecture in Berlin on 16th April 1920. The two started a correspondence in 1921 that lasted for about a year and exchanged the latest spaceflight literature and news between the two countries. Scherschevsky's first article devoted to spaceflight dealt with Tsiolkovsky's priority claim on many original spaceflight concepts but as this appeared in a Russian émigré newspaper, it probably didn't come to the attention of many German readers [3].

Alexander Scherschevsky in 1929

In 1926 he tried to persuade Hermann Oberth's publisher to issue a German translation of Tsiolkovsky's latest works but this was rejected because they felt it contained "nothing new". Thankfully, Scherschevsky made copies of his translation available to interested parties and was responsible for the growing reputation of the Russian pioneer in Germany during the late 1920s [4].

Despite being a relatively prolific writer of articles in the German specialist press, Scherschevsky's only book on the subject was 1929's *Die Rakete für Fahrt und Flug* ('The Rocket for Travel and Flight'). Unfortunately it has been criticised for a lack of historical accuracy, with Willy Ley particularly scathing of a described (but entirely fictitious) face-to-face meeting between Tsiolkovsky, the French rocket pioneer Robert Esnault-Pelterie, and the Russian Tsar! [5].

Willy Ley opined that Scherschevsky should have stuck to writing about his homeland: "He could have done better if he had written about Russia, but he didn't. He knew that nothing much favourable could be said and he would not say anything unfavourable, certainly not in print" [6].

In 1929 Scherschevsky's name came to Hermann Oberth's mind when he was looking for assistants to build a liquid-fuel rocket to launch at the premier of Fritz Lang's science fiction movie *Frau Im Mond* ('The Woman in the Moon'). Against the wishes of the film studio, the director had even agreed to invest 20,000 Marks of his own money into the project. Although Oberth had known of Scherschevsky since 1926, when he finally met him in person he was less than impressed. "I found a Russian emigrant completely stuck in the dirt, and in the real meaning of the word," Oberth later told the President of the *Verein für Raumschiffahrt* (German Spaceflight Society). "I had the impression that if one would throw that guy against a wall he would stick to it! And on top of that, he had no energy and zest for life left, just a totally unproductive gallows humour" [7].

Feeling sorry for him, Oberth hired Scherschevsky alongside ex-pilot Rudolf Nebel on a salary of 6,000 Marks each but he soon regretted the decision. Neither of them had any practical engineering experience and their early amateurish experiments resulted in an explosion that almost blinded Oberth in one eye [8].

Sensing his Russian assistant's only talents lay in the world of writing, Oberth entrusted him with the job of proof reading a new edition of his book *Die Rakete zu den Planetenräumen*. It appears Scherschevsky botched this too because when veteran space engineer Boris Raushenbakh met Oberth in the late 1980s (he was the translator a 1948 Russian edition), he was surprised to hear Scherschevsky being blamed for over one hundred errors in this 1929 edition. Oberth believed Scherschevsky had simply initialled the pages and sent them off to the publisher without reading them [9].

Right from the beginning Oberth had been frustrated by his assistant's apparent lack of drive and poor personal manners. The final straw came when he lent him money to have his bad teeth fixed, only to discover that the Russian had spent it all on sweets [10]. Sacking him, Oberth complained: 'You have, in fact, done no work for me whatsoever' [11]. He later feared the sacking would be seen as anti-Semitic, when in fact it had been caused by the Jewish man's extreme laziness [12].

Surprisingly, Scherschevsky's left-wing sympathies appear to have eluded the staunchly anti-communist Oberth. He even defended him from such accusations by contemporaries: "It is also questionable whether Scherschevsky had already been a Bolshevist or whether Ley and Nebel painted him 'Red' just for their own benefit. Although he has always been full of his (close) relations with the Russian Embassy... because of the whole impression I got of him during the half year of working for me, I believe, that in reality, it was not that deep" [13].

But recent research by Russian scholar Dr. Tanja Jelnina appears to prove Scherschevsky did have close links with the Soviet embassy in Berlin and passed on 32 reports about German rocketry between November 1929 and July 1931. By then the Soviet government had started a recruitment drive to encourage foreign specialists to move to the USSR. Hermann Oberth was approached in February 1932 by a Russian agent named 'Vladimir Kubin' but turned down the offer because he suspected he was being asked to develop missiles for use against Germany [14].

The Soviets then turned to Scherschevsky and persuaded him to return home in the spring of 1932. According to a recent biography of Soviet rocket pioneer Valentin Glushko co-written by his son Aleksandr, the NKVD (*People's Commissariat for Internal Affairs*) recommended the returning émigré for a job at Leningrad's Gas Dynamics Laboratory (GDL) but he was suspected of being a 'plant' placed inside the workshop to keep an eye on them. Bravely, Glushko wrote to the NKVD voicing his objections [15].

Scherschevsky's relations with his new co-workers got off to a bad start when they discovered he was being given full VIP treatment. To their astonishment he was living in a luxury two-bedroom apartment at the Astoria Hotel, ate the best food and dressed in the finest Soviet clothes. He often arrived late at the office, sat at his desk and appeared to fall asleep. When confronted by Glushko, he said he had to close his eyes so the sight of other people didn't disturb his thinking! When Boris Raushenbakh, a Russian of German descent, met this strange new character he was disappointed: "If Scherschevsky was put on a question relating to air travel, the answer would always be the exact identification of the periodical (including even the year and the issue number) in which an article on the given subject appeared, as well as who wrote it, but to get from him even a short summary of the article's content was impossible. His laziness was probably a result of his complete inability to do creative work" [16].

With hindsight, Raushenbakh wondered if the man he met in Leningrad had been changed by some physical or mental illness. "Until 1929 he regularly published articles on the subject of rockets and space in German scientific and popular magazines. These articles…were of good quality and could not possibly have been written by the lazybones Scherschevsky later turned out to be. Either someone else had written them, although it is very unclear who would have done so and for what reason, or shortly before 1929 he went though some sort of illness which affected his capacity for intellectual work. That he had some kind of chronic illness is undeniable: his face was always covered with sores. All in all, he was an interesting companion who loved to make jokes, and who understood what one was talking about as long as the conversation remained general" [17].

During this period, his only contribution to the work of the GDL consisted of some mathematical calculations for a planned meteorological rocket. Its design used a German-style nose-mounted engine to carry a 20 kg payload to an altitude of one hundred kilometres before falling back to Earth via parachute [18].

Scherschevsky even revealed the specifications for Oberth's latest 'Kegelduse' cone-shaped rocket engine but a planned Soviet copy (the ORM-15) was never built. This has been cited as proof of Scherschevsky's lack of influence [19]. When the two main Soviet rockets groups (Leningrad's GDL and Moscow's GIRD) merged in 1933 and moved to the capital under direct government control, Scherschevsky wasn't invited to join. He was quickly forced to give-up his VIP lifestyle and moved into a single room at a communal housing block. Although given a job at a Leningrad college, he was soon translating documents at a local library to make ends meet [20].

Unfortunately, he now had too much knowledge of secret military technology and was trapped inside the Soviet Union just as Stalin's 'Purge' was about to begin. On 7th October 1936 he was at the library with a friend when they were both arrested by the secret police. Although the other man had been their real target, when the NKVD found out who Scherschevsky was they were delighted to discover they had a possible 'German spy' on their hands. During his interrogation he admitted to having "transferred abroad" details of Georgy Langemak and Valentin Glushko's 1935 book *'Rockets: Their Design and Uses'* and was charged with "anti-Soviet activities". On 22nd March 1937, Scherschevsky was given a death sentence and his personal papers were confiscated by

Valentin Glushko's arrest photo

the NKVD. When historian Dr. Tania Jelnina asked to see them in the mid-1990s, she was told they were still under 'lock and key' at the former KGB archives for security reasons. To Glushko's biographers this is evidence that he really had been a "double agent" spying on both the German rocket pioneers in Berlin and the GDL in Leningrad. To quash rumours that Glushko had been responsible for Scherschevsky's arrest, they say that when they asked the St. Petersburg Military Prosecutor to examine the relevant files, no mention was found of Glushko [21].

On 28th May 1937 Scherschevsky was led into a soundproofed cell in the basement of the NKVD's Leningrad 'Bolshoi Dom' (Great House) headquarters and executed. As was normal procedure during the industrialised killing of the Purges, he would have been killed with a single shot to the back of the neck [22].

We now know the persecution of the Soviet rocket group was a direct result of their close links to popular Red Army leader Marshal Mikhail Tukchaevsky. As he had been an early supporter of their work, his patronage proved fatal when he was denounced as the leader of an "anti-Soviet Trotskyite conspiracy" by a jealous and paranoid Stalin [23]. Marshal Tukchaevsky was executed on 12th June 1937 – alongside his mother, sister, and two brothers [24].

Former GDL director Ivan Kleimenov was arrested at the end of 1937 and it is easy to imagine information gathered from Scherschevsky being used against him. He was executed in January 1938. Valentin Glushko's turn for arrest came in March 1938 but by then the purges had run their course and he survived.

Although it is true to say Alexander Scherschevsky's odd personal habits and political naiveté alienated him from his contemporaries and ultimately led to his death, he still deserves to be remembered as one of the first aviation writers to promote spaceflight to the general public. Unfortunately an anonymous death amongst millions of other victims of the 'Great Terror' has effectively purged his name from the space history books.

References:

1. N. Ryan, "Life of A. Shershevskii", in *Interplanetary Flight and Communication*: *Theory of Space Flight*, pp.288-289; "Alexander Boris Scherschevsky", in *Die Rakete*, 3, 42-43, 1929. [I am using the German transliteration of his name but it can also be 'Aleksandr
Borisovich Sherschevskii']
2. Willy Ley, *"Rockets, Missiles, and Space Travel"*, Viking Press, 1958, *p.*127.
3. Tanja Jelnina, 'Dissemination of Information on K.E. Tsiolkovsky's
Scientific Works on Astronautics in the West (Up to the Mid-1930s)', in *History of Rocketry and Astronautics*, Vol 34, Univelt,

2001, 474; Asif Siddiqi, *"The Red Rockets' Glare"*, Cambridge University Press, 2010, p.60.

4. Ibid., Jelnina, p.477.

5. Willy Ley, *"Rockets"*, 1958, p.103.

6. Ibid., p.127.

7. Tanja Jelnina, *"A. B. Scherschewsky - Eine Biograpidsche Skizze"*

(A.B. Scherschevsky – A Biographical Sketch), Oberth Museum monograph 1996.

8. Frank H. Winter, "Fritz Lang's Surprising, Silent Space Travel Classic", *Starlog,* p.66, 1981.

9. B.V. Raushenbakh, *"Hermann Oberth: The Father of Space Flight"*,

West-Art Publishers, 1994, p.66.

10. Ibid., p.67.

11. B.V. Raushenbakh, *"Hermann Oberth"*, 1994, p.67.

12. Mentioned in letter quoted in Jelnina's *"A.B. Scherschewsky"*, 1996.

13. Quoted in Jelnina *"A.B. Scherschewsky"*, 1996.

14. B.V. Raushenbakh, *"Hermann Oberth"*, 1994, p.80.

15. P.I. Kachur and A.V. Glushko, *"Valentin Glushko: Konstruktor Raketnykh Dvigatelei i Kosmicheskikh System"*, St. Petersburg, 2008, pp.105-111.

16. B.V. Raushenbakh, *"Hermann Oberth"*, 1994, p.66.

17. Ibid., p.85.

18. Ye. K. Moshkin, *"Development of Russian Rocket Engine Technology"*, NASA Technical Translation, 1974, p.103.

19. P.I. Kachur and A.V. Glushko, *"Valentin Glushko"*, 2008, p.230.

20. B.V. Raushenbakh, *"Hermann Oberth"*, 1994, p.84.

21. I. Kachur and A.V. Glushko, *"Valentin Glushko"*, 2008, p.230.

22. Igal Halfin, *"Stalinist Confessions"*, University of Pittsburgh Press,

2009, p.172.

23. Robert Conquest, *"The Great Terror"*, Pelican, 1971, p.301.

24. Asif A. Siddiqi, *"Challenge to Apollo: the Soviet Union and the space race, 1945-74"*, NASA, 2000, p.10.

Sir Bernard Lovell and the Sputniks

At the beginning of the Space Age Jodrell Bank in England had a unique status as the only radio telescope capable of detecting signals from the first interplanetary space probes. Both the Americans and Soviets had to go 'cap in hand' to its director Bernard Lovell asking for help.

Ironically, Lovell had become an astronomer by accident. As a physicist he had developed portable radars for British bombers but when he realised they also picked up signals from space an interest in the heavens was born. By the early 1950s he had drawn-up plans for a giant radio telescope capable of detecting cosmic rays and meteors high in the atmosphere, as well as obtaining permission to build it on a University of Manchester-owned farm allotment just outside the city. Despite this official support by the time it neared completion in 1957, it was £250,000 over-budget and his position as director was in serious doubt.

Sputnik would turn out to be a godsend for Jodrell Bank but Lovell was initially reluctant to track the Soviet satellite and it took an urgent call from the British Ministry of Supply to change his mind. When informed that there was no other radar facility in the west able to track the satellite's upper stage rocket (which had also gone into orbit), frantic efforts were made to get the unfinished telescope ready. They were able to track the *Sputnik* upper stage as it passed 125 miles above the English Lake District on 12th October 1957. [1]

The feat made headlines worldwide and Lovell became a media star, with journalists descending on Jodrell Bank asking for his latest opinions on the 'Space Race'. He was even forced to hold two press conferences each evening to accommodate them! On October 26th 1957 Lovell received a telegram from Moscow asking if he could again locate the *Sputnik* rocket as they had lost track of it too. It only took six hours to find it and for his efforts the Jodrell team received a New Year's telegram from the Soviets - a sign that connections had been made behind the Iron Curtain. It wasn't only

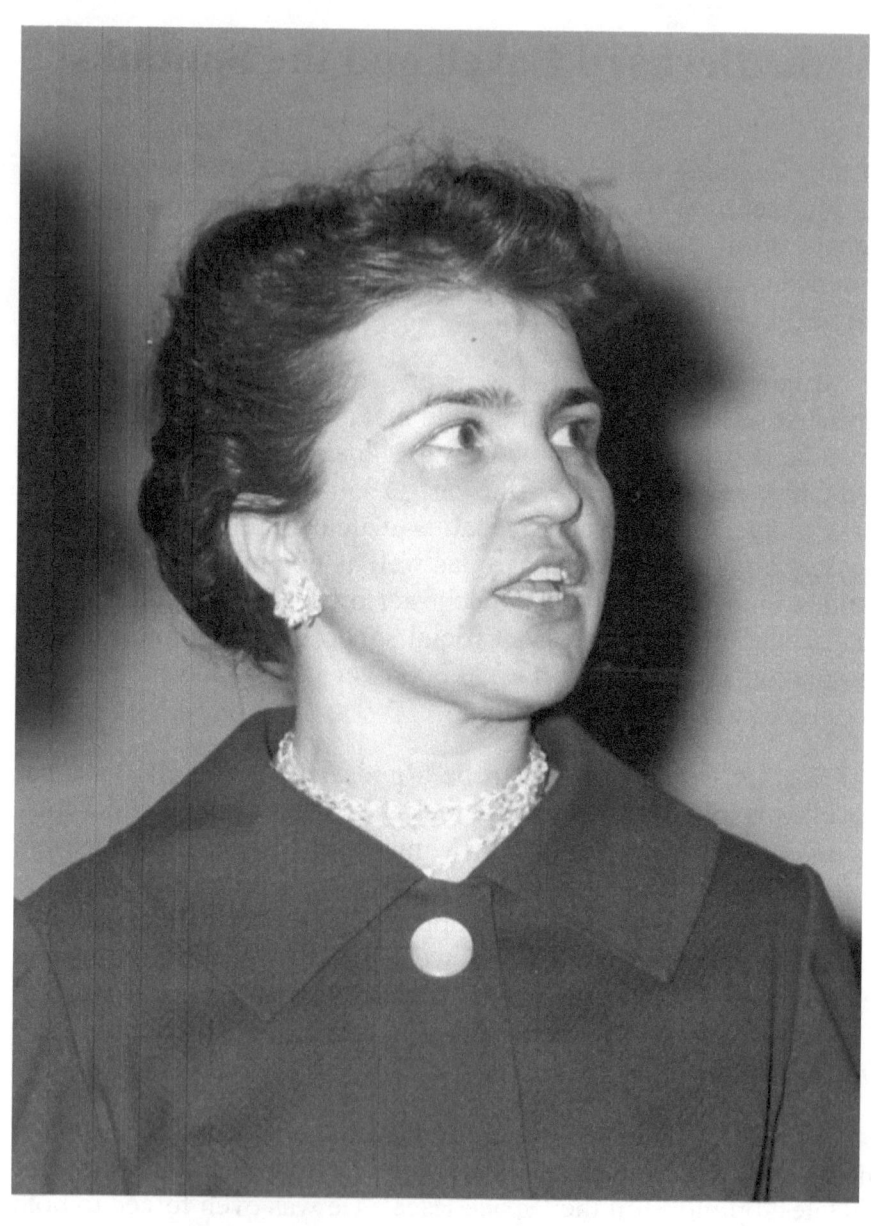

Soviet scientist Alla Massevitch

the Soviets who saw the potential of Jodrell and 1958 saw a high-profile delegation of US Congressmen (guided by future Soviet 'space sleuth' Charles Sheldon) coming to look over the facilities. Although they were impressed by what they saw, they were shocked at its dire financial situation and left promising future American funding. This turned into an hourly rental rate whenever they used it for satellite tracking. This had first happened in early 1958, when Jodrell received a secret message from the US Air Force asking for help with their up-coming *Pioneer* moon missions. When Lovell agreed, American equipment was quietly shipped to England but a *Manchester Guardian* journalist soon spotted large crates on the back of a truck with 'Jodrell Bank, US Air Force, Project Able' written on the side! [2]

Lovell himself was in Moscow for the International Astronomical Union (IAU) congress when this first US Pioneer was launched on August 17th 1958. Unfortunately its *Thor* booster failed and it re-entered the Earth's atmosphere but Lovell used the failure to try coax the Soviets into revealing any similar plans they might have had for a lunar mission. They promised to let him know when they launched one. During this Moscow congress Lovell also established a professional relationship with astrophysicist Alla Massevitch (head of the Soviet satellite optical tracking network), who was a well-known 'ambassador' for Soviet Astronomy. Not only was she a charming and elegant woman but she was also a brilliant scientist who impressed everyone who met her with her flare for foreign languages. [3]

Lovell had first met her several years before when she was part of the official Soviet delegation to the 1955 IAU Dublin congress. When Lovell organised a field trip to England to show off the construction of his radio telescope to astronomers, the Soviets had signed-up to the tour. (Massevitch was also part of a team that returned in June 1961, trying to find their *Venera 1* probe.) When *Luna 1* launched in January 1959, Lovell was annoyed to discover the Soviets didn't let him know as promised and sent a letter of protest to his Moscow contacts. When they followed-up with *Luna 2* in September 1959, Lovell was immediately sent a telegram containing radio frequencies and a predicted lunar impact time but he initially refused to stop a game of Cricket to help. It was only when he was ordered by his wife to respond that he swung into action!

Jodrell was once again packed with excited journalists and they were all impressed when the radio signal stopped at the predicted moment. One can only guess that the Soviets were using Lovell as an unofficial 'record keeper' - someone to verify their achievements to a sceptical west. Right from the beginning Lovell had been passing on information about Soviet rocket stages to the British government but strangely they seem to have been more concerned about the threat of high-flying Soviet bombers and asked Lovell to conduct secret experiments to see if he could track RAF bombers flying at 40,000ft. It has only recently emerged that Jodrell played a part in the west's first 'Ballistic Missile Early Warning System'. President Eisenhower had announced the establishment of such a network in 1960 but construction at its Fylingdales base in Yorkshire was delayed. The military once again turned to Lovell for help and a secret plan codenamed *Verify* saw Jodrell used as a stop-gap during 1962-63. It saw action during the 'Cuban Missile Crisis' of October 1962. [4]

Although Lovell only promised seven minutes warning of a Soviet missile attack, he was assured this was enough time to save a million lives in London and dispatch Britain's nuclear bomber force on a retaliatory strike before their bases were destroyed!

When Lovell was officially invited to the USSR as a 'thank you' for all his help, one can imagine that there was also a faint hope that he could be persuaded to stay. Many of the space facilities Lovell was shown had never been seen by foreigners before and he met prestigious Soviet scientists of the day such as President of the Soviet Academy of Sciences Mstislav Keldysh and nuclear physicist Pyotr Kapitsa. Lovell arrived in Moscow on 25th June 1963 on a scheduled Aeroflot flight. He landed on the same day as a large press conference to celebrate the recent Tereshkova flight and he was taken directly to watch the event but he found it all too 'long-winded'. [5]

The next day Lovell met Academician Keldysh and was told he was to be taken to the then top secret Yevpatoria tracking station in Crimea. During this meeting Lovell also brought-up the topic of a Soviet version of the American Apollo project but wrongly interpreted Keldysh's negative responses as proof that the Soviets were 'out' of the moon race.

During Lovell's visit to Crimea he took part in an astronomy summer school, where he was treated as a distinguished VIP and gave several lectures. On Monday 30th June Lovell was taken to the Yevpatoria tracking station, which he was told was built in a year at a cost of 10-20 million Rubles (£5-10m). Although he wasn't impressed with the overall condition of the facility (he described it as being comparable to the worst British wartime construction), he was taken aback by the solid 'battleship' nature of its dish. We now know its construction started in March 1960 and saw 5,000 naval conscripts working around the clock to finish the complex in just eight months. To save time, Soviet engineers used the hulls of decommissioned submarines for the dish and fashioned its rotation system from the cannon mounting of an unfinished Soviet navy ship. Although he didn't mention it in his contemporary notes of the trip, it was during this visit that a curious incident occurred which Lovell later believed was an attempt to harm him. In a letter to his son released after his death, Lovell relates how he was left alone in a room at the station "for no obvious reason". [6] Much later Lovell came to believe that Yevpatoria's radar beam had been focussed on him to give him a fatal radiation dose! The media naturally had a field day with this pseudo-James Bond story. [7]

A more probable explanation for this awkward incident might have been that his hosts at the tracking station might not have known what to do with him – as it appears he might have been invited by 'higher-ups' who hoped the western scientist might have been tempted to defect to build a 'Soviet Jodrell'. Evidence of this thinking might be seen by the fact that although Lovell had initially travelled to Moscow alone, the Soviets immediately insisted that his wife followed him to Russia for the three week trip. When they eventually hinted to Lovell that they would be able to provide unlimited funds for him to build his dream radio telescope in Russia, Lovell had to reply that as an Englishman he wished to return home as soon as possible. When Keldysh made a reciprocal visit to England several years later, he requested a trip to the Fylingdales radar station with the excuse that Lovell had been allowed to see Yevpatoria but his request was turned down.

From that point on Lovell was anxious to leave Russia. Staying at the *Ukrainia* hotel he lamented that his only source of English news was the British *Daily Worker* newspaper. To add to his stress levels, he discovered that his hotel was full of female delegates attending a 'World Federation of Women' congress and this strange atmosphere wasn't helped when some of the American delegates started singing 4th July songs to mark their Independence Day. The next day he took the overnight train to Leningrad to see some of that city's cultural sights before he flew to Armenia to tour another observatory. During his last two days in Moscow Lovell tried to follow-up on an earlier invitation to visit Pyotr Kapitza's Dacha. After some stalling by Massevitch, Lovell was taken on his last day and had a brief chance to talk to Kapitza about the space programme. Although the Russian admitted he "knew too much" to talk freely he did reveal that the Soviets hoped to send a cosmonaut around the Moon in 1967. It is not surprising that Lovell later admitted he only felt safe when his plane finally touched down in London.

Lovell's relationship with the Soviets appears to have cooled after his Moscow trip - the final straw being the infamous 'scoop' of February 1966 when he published pictures from the lunar surface taken by *Luna 9* before the Soviets did. This happened because his team recognised the probe's signal as being similar to a press agency wire photo signal. A machine was quickly obtained from the *Daily Express* Manchester office and they were all astonished to see the first picture from another world emerge when it was hooked-up to the telescope. Lovell naturally assumed that the Soviets would reveal their version of the photograph at a press conference then taking place in Moscow and he gave his copy to the journalists at Jodrell. Unfortunately the Soviets hadn't unveiled their own picture straight away and were less than amused to see

The front page that enraged the Soviets

their moon pictures published in the west first. Many years later prominent Soviet planetary scientist Roald Sagdeev revealed that this incident tarnished Lovell's reputation and he lost membership of the Soviet Academy of Sciences because of his indiscretion. [8]

By that time the Soviets had their own tracking stations capable of matching Jodrell and had stopped asking Lovell for direct help. His swansong came during the monitoring the high-profile (but mysterious) *Luna 15* mission in July 1969. As they listened, signals from the probe stopped suddenly on the evening of 21st July when the unmanned lunar lander crashed onto the moon and was destroyed. As always Bernard Lovell was on hand to provide quotes for the next day's newspapers! [9]

References:

[1] *Space Has No Frontier: The life of Sir Bernard Lovell*, John Bromley-Davenport, pg. 192, Bene Factum Publishing 2013.

[2] *Bernard Lovell: A Biography*, Dudley Saward, pg. 205, Robert Hale 1984.

[3] *Begegnungen eines Raumfahrt-Enthusiasten*, K. H. Marek, Verlag Iniplu 2000. (Thanks to Jacqueline Myrrhe for Massevitch chapter translation)

[4] 'Diversions of a Radio Telescope', F. Graham-Smith & B. Lovell, *Notes of the Royal Society*, pg. 197-204, 2008.

[5] Lovell notes, John Rylands Library, University of Manchester.

[6] Bryan Lovell letter, Rylands Library.

[7] 'Sir Bernard Lovell claims Russians tried to kill him with radiation', *Daily Telegraph*, 22 May 2009.

[8] *Space Has No Frontier*, pg. 233.

[9] *Soviet and Russian Lunar Exploration*, Brian Harvey, pg. 214, Springer Praxis 2007.

Reporting the Right Stuff?

During the Cold War the Soviet Union did everything possible to control press coverage of its space programme. Unfortunately for the Kremlin over fifty Moscow-based foreign correspondents were able to 'scoop' some of its carefully stage-managed cosmonaut propaganda.

A 27-year-old Nicholas Daniloff arrived Moscow in 1961 to work for legendary *United Press International* (*UPI*) bureau chief Henry Shapiro – a veteran reporter who had lived in the Soviet Union since the 1930s and had excellent contacts. Although Daniloff was born in Paris and raised in the United States, he was from an aristocratic 'White Russian' family and had a genuine interest in the culture. Unfortunately he found his new assignment mainly consisted of busy days in the office monitoring Soviet newspapers and television for items to rewrite as *UPI* 'wire copy'. He estimates 75-80% of western reports from Moscow during the 1960s originated in the Soviet media. Correspondents who wanted to make a reputation for themselves had to get out of the office and find independent sources – a rare commodity during the Cold War. The fact that Nicholas Daniloff's grandfather had been a wartime general on Tsar Nicholas II's staff didn't make this any easier.
"On Mondays, Wednesdays and Fridays, I was seen as hostile. On Tuesdays, Thursdays and Saturdays, I was seen as friendly or at least understanding!" he half-joked [1].

Ironically, Soviet Premier Nikita Khrushchev offered a rare glimpse behind the official façade because his own enthusiasm for spaceflight - one of his few propaganda successes against the west - made him especially talkative on the subject. Western reporters literally stalked him around Moscow, notebooks in hand, waiting for a revealing quote.

"Spaceflight was very important to Khrushchev because it made the USSR look like a very modern, technological state - which was a misperception," recalled Daniloff. "The USSR gave space a high priority - it was a 'Guns over Butter' sort of thing."

For decades press dispatches had been directly censored at the central telegraph office before transmission to the west. In early 1961 Khrushchev allowed news bureaus install their own private 'telex' machines, so they suddenly had uncensored communications with the outside world [2].

In the run-up to Yuri Gagarin's historic spaceflight on 12th April 1961, Moscow was so 'electrified' with rumours of the flight that western journalists heard them. *UPI*'s Henry Shapiro was even confidently reporting on 10th April that his own 'reliable but unofficial sources were telling him it had already taken place [3].

A similar report by Dennis Ogden of the British communist newspaper *The Daily Worker* caused the Soviets the most trouble. Although he should have needed little controlling - he regularly played cricket with exiled spy Guy Burgess! – his story was published on 12th April and wasn't just for western consumption. As his newspaper was one of the few western publications allowed on sale in Moscow, it appeared on newsstands around the city on the very morning Gagarin was launched. Soviet journalist Leonid Finkelstein remembered seeing the headline and was in the process of checking it out with his own sources when the first news of Gagarin broke. He later learned those responsible for allowing *The Daily Worker* on sale were sacked, whilst Ogden had to leave the country the following year [4].

BBC correspondent Reginald Turnill attended Gagarin's first press conference but felt the event was solely organised to humiliate the western press. A delegation of Soviet nurses was paraded past them and they then had to listen to them all laughing at the mocking answers given to their questions [5].

Soviet journalists also caused problems for the authorities. Astronomer Alla Massevich had been drafted in to monitor their output but her own ignorance of the fact that Gagarin's landing by personal parachute was top secret information led her to approve its use in several newspapers. She was soon relieved of her censoring job and full details of the landing weren't revealed for another ten years [6].

When cosmonaut Andrian Nikolayev was launched aboard Vostok 3 in August 1962, western correspondents were astonished to hear their sources tell them a second cosmonaut would be launched the following day. As this simply seemed impossible, they dismissed it as 'Soviet banter' (or a joke) and missed the scoop [7].

"In the 1960s, and later, fear of *KGB* harassment and repression, were a constant element of Soviet life," recalled Daniloff. "There were no scientific sources other than official press conferences. Soviet newsmen, especially at *TASS*, got advance notice of spaceflights and would pass on some info but only occasionally. I had a source at the *Novosti* press agency who gave me some stuff about Tereshkova being unlikely to fly again [8]."

He filed his 1964 exclusive about Valentina Tereshkova being medically disqualified after a Caesarean but later discovered his source was a *KGB* man feeding him harmless stories to gain his trust. Luckily, their friendship had cooled after the man tried to pressure him into revealing the source of another more sensitive political story. Anyone doubting the compromised position a correspondent could find himself in only needs reminding that Daniloff was arrested during a second 'tour of duty' in the 1980s and only released as part of a classic Cold War spy swap!

Another source suspected by the western press pack of having *KGB* connections was Viktor Luis. Although employed as a 'stringer' by many news outlets in Moscow, he was widely believed by them to be used to leak information the Soviets wanted known. These *KGB* sources were disparagingly termed 'Nursemaids' (*Nyanka*) by western correspondents.

In mid-March 1965 Henry Shapiro rushed into the *UPI* office after one of his sources (his assistants knew never to ask who) told him a spaceflight was days away and would see one of the cosmonauts conduct the first spacewalk. Although details were hazy, the source revealed it was a two-man crew and one of them would leave the spacecraft using a "special capsule" – which we now know was probably a reference to Voskhod's unique throw-away inflatable airlock. Daniloff wrote-up the story and transmitted the *UPI* scoop to the west 24-hours before Alexei Leonov and Pavel Belyayev were launched [9].

After their historic mission the *Novosti* press agency approached *NBC* television correspondent Frank Bourgholtzer with an offer of an exclusive interview with the Voskhod 2 cosmonauts.

A New York TV crew were dispatched to record it and even asked amateur artist Leonov to illustrate the broadcast with his own drawings. When Leonov asked them to bring along some hard to find 'Magic Marker' pens, the Americans arrived with six dozen for the surprised cosmonaut [10].

Perhaps unfamiliar with western interview techniques, the cosmonauts convinced themselves they were really talking to a secret delegation sent by *NASA*. In his 2004 autobiography *Two Sides of the Moon* Leonov repeats the claim [11].

"Our meetings with the delegation took place in the offices of a Soviet press agency", he wrote. "At first I thought the delegates were American journalists but I soon realised that they were not, that they were specialists from NASA…They sat asking questions for hours and hours. They recorded the interviews, and filmed them with several cameras under special lighting. I do not believe that what NASA learnt from us during those discussions was put in a drawer and forgotten."

When *NBC*'s 'The Man Who Walked in Space' was broadcast on American television on 14th May 1965 it was watched by millions - including astronaut Ed White, who admitted footage of Leonov's spacewalk helped him learn how to float around in orbit [12].

The Kremlin tried to isolate Moscow-based correspondents by only announcing the first details of new missions via its *TASS* office in London. By the time details filtered back to them, their editors had already published without waiting for their expert opinions. Often the only advantage they had was an ability to read ten Soviet daily newspapers and as many 'provincials' as they could get their hands on. Stories that might have been censored in the capital often appeared in these obscure publications. Henry Shapiro delegated this task to his Russian wife Ludmilla, who spotted a reference to the first multi-man Voskhod 1 crew in a provincial newspaper before it flew [13].

Ironically, she had a real-life 'Deep Throat' source inside the Soviet space establishment as her cousin was Yuri Pobedonostsev – a rocket engineer who had been sent to Germany after the war to study captured V2s. She later admitted he told her nothing for fear of revealing state secrets.

According the Daniloff, the correspondent who best exemplified the classic 'beat reporter' in Moscow was Theodore Shabad of *The New York Times*. Born in Berlin in 1922, he moved to America aged 16 but had also learned Russian to keep in contact with his father in the Red Army. Shabad was a trained geologist – something colleagues believed helped him find good academic sources [14].

"He was not a trained journalist (but) worked on the *Times* copy desk before going to Moscow," remembered Daniloff. "He made an effort to connect with Russians and attended dissertation defences at Moscow University and elsewhere."

Shabad's biggest 'scoop' came after attending the wedding reception of cosmonauts Tereshkova and Nikolayev in November 1963. Told by Russian friends that two of the guests were 'top secret' rocket engineers, Shabad learned they were named Sergei Korolev and Valentin Glushko. He obtained file photos under those names from *TASS* and wrote his exclusive two years before the Soviets officially admitted Korolev's existence [15].

Sometimes the Soviets had to be embarrassed into admitting a western news story. This happened during the homecoming parade for the Soyuz 4/5 cosmonauts in January 1969, when a lone gunman opened fire on the motorcade in an assassination attempt on Premier Leonid Brezhnev. Although a driver was killed and several cosmonauts injured, it took repeated questions from the foreign press pack before the Kremlin finally admitted it happened [16].

Unlike the western press, Soviet journalism played a central role in the Soviet propaganda machine. Official space correspondents like Alexander Romanov, Evgeny Riabchikov and Yaroslav Golovanov might have been given unprecedented access to a top secret military project but they always knew they were simply there to portray cosmonauts as living examples of Marxist-Leninist ideology – even if they knew otherwise. Their articles were censored at a special office on Moscow's Molodyozhnaya Street but it became a bureaucratic nightmare involving the submission of multiple forms and copies of articles [17].

Youth newspaper *Komsomolskaya Pravda* space correspondent Yaroslav Golovanov had even given-up a promising rocket engineering career for journalism to be closer to his heroes.

Although his twenty-two million readers ensured close access to all the leading space personalities, he soon grew disillusioned.

"It was very difficult for me to work because I knew much more than I was able to write," he admitted in 1990. "Soviet journalism is journalism between the lines. If you could not say something directly, you could still write for clever readers [18]."

Before a manned mission Soviet pressmen were given special envelopes containing the photographs and biographical details of the new crew but weren't allowed to open them until told. This didn't stop them telling western friends when the packages had arrived on their desks [19].

Lars Bringert of the Swedish newspaper *Dagens Nyheter* got these tip-offs and passed them onto others in the western press [20].

American Joseph Galloway found a way to verifying these tip-offs when he discovered a Russian friend had a neighbour involved in manned rocket launches. When this friend was asked by the neighbour to collect his mail, Galloway correctly guessed the engineer had flown down to Baikonur for a launch [21].

One of the few Soviet pressmen to escape censorship was Leonid Finkelstein. He defected during an official visit to London in 1966 but discovered nobody believed his insider view of the Soviet space programme. Westerners simply didn't believe the United States could beat the USSR to the Moon and his book *The Russian Space Bluff* (1971) wasn't published until after the triumph of Apollo 11. He remained in Britain under the pseudonym 'Leonid Vladimirov' and became a well-respected broadcaster on *BBC World Service* radio.

Nicholas Daniloff swapped Moscow for Washington D.C. in 1965 and also decided to write a book on Soviet spaceflight. Between 1968 and 1972 he researched it part-time at the Library of Congress, interviewing émigrés and space experts such as Frederick Durant and Charles S. Sheldon.

"When you are in Washington, of course, the slant is going to be US positive," said Daniloff. "I did on one occasion go to the *CIA* for a briefing on Soviet space activities. Needless to say, nothing classified was alluded to [22]."

The Kremlin and the Cosmos was published in 1972 and sold

5,000 copies. "The Soviets later said the book was faulty and lacking in correct information. Not surprising! I did the best I could to scoop up info but obviously I was an outsider looking in."

By the 1980s things were changing rapidly for the press behind the Iron Curtain. Previously off-limits topics were open to scrutiny; whilst Mikhail Gorbachev's policy of *glasnost* allowed Soviet journalism to publish all the 'space secrets' it had known for decades. The close relationship between the Soviet press and the Kremlin also nearly paid-off with a 'Journalist in Space' mission to the Mir space station. In the mid-1960s, Yaroslav Golovanov had been selected alongside broadcaster Yuri Letunov and *Krasnaya Zvezda* writer Mikhail Rebrov for a similar propaganda flight.

That plan had come directly from Chief Designer Korolev, who was disappointed by the bland descriptions given by the cosmonauts and wanted a professional communicator aboard one of his spacecraft. Unfortunately the idea was scrapped after his sudden death in 1966 [23].

Twenty years later the Soviet press were outraged again when the new commercial space agency *Glavkosmos* offered a 'space tourist' seat to a Japanese journalist. The *USSR Union of Journalists* organised its own cosmonaut selection process to embarrass the Kremlin into supporting a similar flight by one of its members. It even sent an official delegation to Baikonur to watch the launch of Soyuz TM-8 in September 1989.

Pravda editors asked Mikhail Gorbachev who would fly first but all he could do was diplomatically remind them that the Japanese were already in training. Tellingly, a readers' poll in the newspaper showed 70 percent were opposed to sending the Soviet journalist. Six finalists were announced in May 1990 but the project fizzled out for lack of funds [24].

"To cut a long story short, there has been much talk – but business, as usual, was done by the Japanese," is how Golovanov lamented the final act in the story of Soviet Space Journalism.

******

1. Nicholas Daniloff to author 29th July 2016.

2. Nicholas Daniloff, *Of Spies and Spokesmen*, Univ. Missouri Press, 2008, pg.91.

3. Henry Shapiro (*UPI*), "Man in Space Report Electrifies Russia", *Vancouver Sun*, 10th April 1961.

4. Leonid Vladimirov, *The Russian Space Bluff*, Tom Stacey Ltd. 1971, pg.102.

5. "Space Miscellany", *Spaceflight*, **21**, pg.382, 1979.

6. Slava Gerovitch, *Soviet Space Mythologies*, Univ. Pittsburgh Press, 2015, pg.94.

7. Reuters, "Soviet Launching 'Joke' Was Lost on Newsmen", *The New York Times*, 14th August 1962.

8. Daniloff interview July 2016

9. Nicholas Daniloff, *Of Spies and Spokesmen*, Univ. Missouri Press, 2008, pg.121.

10. Whitman Bassow, *The Moscow Correspondents*, Morrow & Co, 1988, pg. 212.

11. David Scott & Alexei Leonov, *Two Sides of the Moon*, Simon & Schuster, 2004, pg. 122.

12. Glenn E. Schweitzer, *Techno-Diplomacy*, 1989, p.168.

13. Bassow, *The Moscow Correspondents*, pg.253.

14. "Theodore Shabad, A Times Editor and Geographer, is Dead at 65", *The New York Times*, 6th May 1987

15. Nicholas Daniloff, *The Kremlin and the Cosmos,* Alfred Knopf 1972 pg.69

16. Daniloff, *The Kremlin and the Cosmos,* pg.199

17. Vladimirov, *The Russian Space Bluff*, pg.104.

18. "What's In the Stars for Soviets", *The Chicago Tribune*, 13th December 1990.

19. Bassow, *The Moscow Correspondents*, pg.253.

20. Nicholas Daniloff, *Of Spies and Spokesmen*, Univ. Missouri Press, 2008, pg.111.

21. Bassow, *The Moscow Correspondents*, pg.355.

22. Daniloff interview July 2016

23. Philip Clark, *The Soviet Manned Space Programme*, Salamander 1988, pg.30.

24. Gordon Hooper, "The Year of the Journalist", *Spaceflight News*, pg.18, May 1990.

The Eagle and the Bear

In Europe only citizens of the Soviet Union and isolated Albania missed one of history's most inspiring moments as most of the Eastern bloc had decided to ignore the Kremlin's wishes and broadcast live coverage of Neil Armstrong and Edwin "Buzz" Aldrin landing on the moon. A privileged few in Moscow did get caught up in this 'Apollo Fever' as the military and space elite assembled at a secret monitoring centre on Komsomolsky Avenue to follow the mission.

Unlike others in attendance, however, cosmonaut Alexei Leonov's reactions were tempered by the knowledge that if things had worked out a little differently, it could have been him and not Neil Armstrong who became the first person to leave a footprint in the lunar dust. Four years earlier he had become famous throughout the world as the first person in history to walk in space, but since then he had been working toward a new and even more impressive goal. As he witnessed those flickering images emanating from the surface of the moon, he would undoubtedly have experienced some frustration. And yet, at the same time, he felt privileged to be seeing history unfold before his eyes.

"It was with mixed emotions that I stood watching events unfold on our television monitors that July morning," Leonov admitted years later. "If it couldn't be me, let it be this crew, I thought, with what we in Russia call 'white envy' – envy mixed with admiration."

Although Leonov realised he and his nation had lost the race to be first, he still believed – like many in the room – that Soviet cosmonauts would follow the Americans to the moon. As their conversation turned to professional talk about the mission, one of the military officers present sympathetically patted Leonov on the back with the words, "That's how it's done...that's the task that lies ahead of you."

Although the moon race might have been important in the early 1960s, by the end of the decade the White House and the Kremlin had other things on their minds. The Nixon administration was preoccupied with ending the Vietnam War, while Moscow was busy trying to re-establish Soviet authority over an increasingly rebellious Eastern Europe as well as trying to prevent a war with China over a disputed border.

After the Apollo 11 moon walk the Kremlin grudgingly acknowledge the United States' feat by devoting seven minutes to the mission on the main evening news, while the *Pravda* newspaper relented on its previous indifference in providing a six-column account of the successful landing. Unfortunately for Alexei Leonov, this recognition, however restrained, also signalled the Kremlin's own loss of interest in the moon race, having been convincingly beaten in the eyes of the world. The party line now was that there had never been any thoughts of trying to send Soviet cosmonauts to the moon, not when unmanned probes could do a better job for far less risk.

In retrospect, any chance of a Soviet cosmonaut walking on the Moon died along with legendary 'Chief Designer' Sergei Korolev on a Moscow operating table in January 1966. His OKB-1 (Special Design Bureau Number 1) in Kaliningrad, northwest of Moscow, had played a pivotal role in the space program's success up to that point. Unfortunately, the man selected to replace him as head of OKB's sixty thousand workers, his loyal deputy Vasily Mishin, simply was not up to the job.

Mishin's career in rocketry had begun at the end of WWII after he was dispatched to Germany to study the V-2 rocket. Back in Moscow in 1946 he became Korolev's deputy and was deeply involved in the launch of Sputnik. Unfortunately he displayed little of Korolev's uncompromising drive, initiative and foresight and only seems to have gained the top job because he was a compromise candidate acceptable to all.

Although many imagined the Soviet space program to be one vast monolithic enterprise bringing communist-style central planning to the cosmos, we now know that in reality there were a number of design bureaus literally fighting amongst themselves for the patronage of the military and the Kremlin. Any miscalculation was quickly used by a rival, resulting in a bitter game of personal attacks that seriously hindered the manned lunar effort.

"Without Korolev, Mishin was lost," believed cosmonaut Alexei Leonov. "He was a very good engineer, but he had his weak points; one of which was that he drank. He was also hesitant, poor at making decisions and reluctant to take risks. This was to cost us dear."

Mishin's main rival was Vladimir Chelomey's OKB-52 design bureau based in the Moscow suburb of Fili. Unlike Mishin, Chelomey was a sophisticated, well-dressed man who had already cultivated excellent relations within the military. Although he only had a staff of 8,000, the fact that Chelomey had hired Khrushchev's son to work on missile designs opened many doors for him in the early 1960s. Chelmoley even changed the name of his OKB-52 bureau to the near-identical Central Design Bureau of Machine Building, (TsKBM)!

Mishin also had a tense relationship with General Nikolai Kamanin, the head of cosmonaut training. Since 1959 cosmonaut training had been an all-military matter but by the mid-1960s Mishin wanted his own civilian engineers flying on his design bureau's spacecraft. Unfortunately Kamanin was a strict Stalinist who ran the cosmonaut centre like his own private fiefdom, and the two soon argued over whether these new cosmonauts should be trained directly at TsKBEM or in the existing cosmonaut training facility that later came to be known as Zvezdny Gorodok (Star City). To Kamanin, the fact that Mishin was also a heavy drinker compounded the problem, as he often had to censure or even ground cosmonaut candidates when they got into alcohol-fueled trouble.

"Mishin wants us to regard him as a resolute leader as well and tries to copy Korolev but not having the authority, the experience and the knowledge of Korolev, he often ends up in unpleasant situations," Kamanin confided in his private diary. His once-secret journals offer a unique insider account of the period. Although published in the late 1990s following the collapse of the USSR, they only gained a wider audience in the west after a detailed analysis by

Russian-language speaker and space sleuth Bart Hendrickx.

Sergei Korolev had originally considered sending a Vostok capsule around the moon, but he soon realized that such a complex mission was beyond the capabilities of both its simple design and a solo pilot. He then started work on what we now know as the Soyuz spacecraft, to carry three cosmonauts on a circumlunar mission by the middle of the 1960s. It was only after the full, ambitious details of the NASA lunar landing plan became public that he decided to scrap this small-scale thinking and match the grand scale of Apollo.

The original concept for the giant N1 booster had been on the drawing board since 1956, envisaged as a massive booster capable of launching military satellites, interplanetary probes and even a large space station. To do so, Korolev had to rely on Valentin Glushko - the only man who could have provided the powerful rocket engines he needed for the proposed super booster. Glushko and Korolev had both come of age during the early days of amateur rocketry before they were swept up in the brutal Stalinist purges which nearly destroyed the country in the late 1930s. Both men managed to survive, although they remained deeply suspicious of each other's involvement in their own, unwarranted arrest by the brutal NKVD secret police. Following World War Two they were both sent to Germany to learn the secrets of Wernher von Braun's V-2 rockets, returning to establish their own missile design bureaus on Stalin's orders. Their relationship mired in a long-standing mistrust of each other. A further feud developed when Korolev was selected as the Chief Designer of the main OKB-1 design bureau outside of Moscow, with Glushko's own OKB 456 factory in a subservient role, supplying rocket engines.

Glushko's ill-feeling centered on the not-unreasonable attitude that his engines were the most important part of any rocket. He once boasted that a wooden stick would fly with one bolted on. Korolev, meanwhile, believed the engines were only a part of his overall creation. In years to come, their failure to reach a compromise would doom the Soviet moon project.

By the early 1960s, Glushko's renamed Energomash design bureau was championing the hypergolic nitrogen tetroxide/UDMH propellant mix, originally developed for use on silo-stored ballistic missiles that needed to be launched at a moment's notice. Korolev had an aversion to using these highly-toxic propellants for flights into space, fearing that any accident would result in a catastrophic explosion similar to an R-16 missile at Baikonur in October 1960, which had killed over 120 army personnel and rocket engineers when it exploded on the pad. The fuel had subsequently become known as the 'Devil's Venom' and Korolev was determined to prevent its use on any manned spaceflight. Instead he insisted on the more familiar liquid oxygen and kerosene engines.

That devastating 1960 explosion had strengthened Korolev's hand against yet another rival, Mikhail Yangel's OKB-586 design bureau. Yangel had proposed his own giant moon rocket called the R-56 but was soon told to forget about manned spaceflight and concentrate on improving his missiles. Now it was Glushko who had the stronger hand. He called Korolev's bluff by walking away from the N1 project, making life exceedingly difficult for his main rival.

When Soviet rocket engineers first learned that NASA wanted to use five massive first stage engines and liquid hydrogen/liquid oxygen upper stages on their Saturn V moon rocket, they had serious doubts that they could perfect the technology by the end of the decade. But Korolev soon realized he had been wrong to underestimate the Americans. That coupled with Glushko's refusal to develop the needed N1 engines forced Korolev to turn to aircraft jet engine designer Nikolai Kuznetsov to develop a new liquid oxygen/kerosene engine for the moon rocket. Although Kuznetsov lacked experience with rocket engines, Korolev gambled that by clustering enough of them together, he could create a giant booster equal to the Saturn V. The overall thrust would have to be a fifth more; making it the most powerful booster in the world, and its first-stage engines alone would consume some forty tons of propellant just to get the 345-foot booster off the pad.

Unlike the three-stage Saturn V, the N1 used liquid oxygen and kerosene on all its five stages. After the three conical stages were used to get the lunar complex in low Earth orbit, the fourth stage would begin the Translunar Injection burn to send it on its way to the moon. A fifth stage would perform mid-course corrections, the Lunar Orbit Insertion and the final burn to drop the lunar lander out of its lunar orbit at the beginning of the moon landing decent.

Although it could only place about ninety-five tons into a 51-degree inclination Earth orbit when launched from Baikonur, it is still startling to realize that if the N1 had been launched from the better 28-degree latitude geographic location of Florida it could have placed nearly 105 tons into orbit - a respectable figure compared to its far more sophisticated American rival.

Unfortunately this brute force approach would need a total of thirty NK-15 engines on the first stage, eight NK-15s on the second stage and four NK-19s on the third stage just to get it into Earth orbit. In fact, so many engines were being used that Kuznetsov's engineers decided to downgrade the performance of each engine in order to lower the chances of failures during a launch. To save time only two out of every six engines leaving the factory could be tested, so by extension just ten were guaranteed to work at ignition. Somehow they hoped the mighty booster could be "tamed" in flight and were resting their hopes on the primitive KORD computer system to shut down any faulty engines during the launch. This amazingly cavalier approach was compounded by a decision to ignore testing of a complete first stage on the ground in order to concentrate scarce resources on the actual flight hardware. In contrast to the Americans, who had conducted their first full-on testing of the Saturn V's first stage engines in mid-1965, the Soviets wouldn't get to do the same until they ignited all the engines of an actual N1 during its maiden flight It was an appallingly risky undertaking, and even some of its own designers expected around ten N1 rocket failures before a successful flight was achieved.

As senior N1 engineer Vladimir Vakhnichenko later admitted, "The underestimation of the scale factor - the immense size of the launch vehicle, each launch of which was an event in the life of the country - played a fatal role. It was no big deal that for some rockets it was necessary to carry out forty to fifty launchings before they 'learned' to fly, but that approach was unsuitable for the N1."

Not surprisingly, some of Korolev's own OKB-1 staff had been openly skeptical of the whole project even before it left the drawing board. One of the most vocal critics was a highly-experienced engineer named Konstantin Feoktistov. "The work started in a more or less serious way in the spring of 1964," he recalled in an interview with Bert Vis in Washington, DC, in 1992. "This was the beginning of some very severe contradictions between me and Korolev. I felt the lift capacity of the N1 rocket was not sufficient for a lunar expedition and I thought that it wouldn't work." Korolev appears to have gotten around Feoktistov's objections by the simple expediency of offering the ambitious young man a much-coveted seat on the newly-built, three-man spacecraft called Voskhod.

"This enabled him to get me away from this program for a couple of months so that he could use my engineers by getting me out of the picture," Feoktistov reflected. "After coming back from the flight I went back to the design bureau and had to keep working on the lunar program but…I kept saying that it wasn't going to work and that the project would die sooner or later."

Soviet leader Khrushchev himself was clearly also losing his early fascination with Korolev by 1964 and decided to allow Chelomey into this bitter moon mix after the designer had proposed a two-man circumlunar spaceship called LK-1 (which resembled a cross between a Gemini and Apollo capsule) for launch on the new UR-500 booster, which later came to be known as the Proton rocket. That August, Chelomey took over the circumlunar project altogether so that Korolev could concentrate on the troubled N1-L3 moon landing plan. Unfortunately Korolev was intensely jealous of Chelomey, so when Khrushchev was unexpectedly deposed by the Leonid Brezhnev faction in the Kremlin in October 1964, the Chief Designer put pressure on the new leaders to allow him to resume control of both projects. Chelomey was ordered to scrap his LK-1 plan and instead launch a deconstructed Soyuz on a 'lunar loop' using his UR-500 booster. He was also given a timetable: the flight would be ready in time for the fiftieth anniversary of the October Revolution in 1967.

Even General Kamanin was surprised by all this political maneuvering, complaining that "Korolev scored an easy victory over Chelomey, but this easy victory cost us almost two years."

The new L1 craft was basically a Soyuz minus its spacious front orbital module. It would carry a crew of two cosmonauts without spacesuits on a week long free-return loop around the moon, followed by a safe touchdown back in Russia. As it had a new, thicker heat shield to protect it from the higher lunar velocity re-entry, the craft's designers insisted on several unmanned test flights before any cosmonauts were permitted to occupy the spacecraft on an actual mission. Eventually fourteen complete L1s were built, of which three could have been used for crewed missions.

Considering his earlier bitter opposition to hypergolic propellants, Korolev's agreement to use a Proton vehicle to launch the lunar Soyuz was surprising, as the UR-500 was an old ICBM (Inter-Continental Ballistic Missile) design using the very same toxic storable propellants over which he had fallen out with Glushko. When it came to the N1-L3 moon landing plan, the Soviets decided to copy the American LOR, or Lunar Orbital Rendezvous approach. Ironically this plan - grudgingly accepted by NASA after realizing it was more economical than its favored options of EOR (Earth Orbital Rendezvous) or Direct Ascent - was inspired by the writings of Ukrainian space theorist Alexander Shargey. He had been forced to live under the pseudonym of Yuri Kondratyuk to avoid the attention of the NKVD during the 1930s, before being killed fighting the invading Germans in 1942.

After the disappointment of losing the circumlunar project, Chelomey had asked the Kremlin to adopt his own moon landing plan to replace the troubled N1-L3 at the time of Korolev's death. This LK-700 plan envisioned a forty-five-ton Direct Ascent lunar lander launched on a UR-700 booster - a new giant booster made up of clusters of Proton propellant tanks fitted with nitrogen tetroxide/UDMH rocket engines that Glushko had earlier offered Korolev for use on the N1. One of its strongest supporters, General Kamanin, even claimed the UR-700 would have cost one tenth as much to develop as it was based on Proton components and could be launched from the existing N1 launch pad at the Baikonur Cosmodrome.

The seriousness of his proposal lead to the formation of the Keldysh Commission in late 1966, headed by the respected Academy of Sciences President Mstislav Keldysh, which examined the merits of both moon landing concepts. Not unsurprisingly, as many of the commission's thirty-six members were still sympathetic to the memory of the recently deceased Chief Designer Korolev, they decided to continue with N1-L3.

Since work had visibly begun on the launch pads for the N1 moon rocket, American spy satellites had been keeping watch for any concrete evidence that the Soviets were still competitors in the moon race. Their concerns were somewhat premature; even the launch pad was behind schedule. Vladimir Barmin, the man who had designed the pads for all of Korolev's previous rockets, initially refused to work on the N1 before being reluctantly persuaded.

The new assembly building for the booster measured some sixty meters tall by 190 meters wide and 240 meters long, and would be the largest concrete hangar in Europe or Asia. When the CIA first saw satellite imagery of the construction they knew it was for something special, as it closely resembled NASA's Vehicle Assembly Building (VAB) at the Kennedy Space Center, albeit laid on its side. Barmin's team had decided to scale-up the horizontal assembly methods used on smaller rockets so they wouldn't waste concrete constructing a tall building like the one in Florida. Unlike the Saturn V, which would exit the VAB mounted vertically above a gigantic tractor crawler, the N1 was designed to be loaded horizontally onto flatbeds and hauled by locomotives to the launch pad. The decision had also been made to build the two N1 launch pads less than 2,000 feet apart to save money – an economy measure they would later regret.

Construction work on the N1 itself started in mid-1965 when two hundred factories in Samara, a vast city in the south-eastern part of European Russia, began making components. These were then taken by rail to the Baikonur Cosmodrome in Kazakhstan for assembly. While the Saturn V used cylindrical propellant tanks, the N1's conical design was the result of a more utilitarian approach using progressively smaller internal spherical fuel tanks in each stage to maximize the volume while minimizing construction materials.

Although a Saturn V mock-up was taken to a Florida launch pad as early as May 1966, it would be another year and a half before an American spy satellite managed to photograph its counterpart at Baikonur. By then the CIA was confidently predicting that a Soviet moon landing couldn't take place before 1972, but NASA could still be upstaged by a cosmonaut looping around the moon before Kennedy's deadline could be achieved.

Alexei Leonov had been selected to command the L1 group. "Vasily Mishin's cautious plan called for three circumlunar missions to be carried out with three different two-man crews, one of which would then be chosen to make the first lunar landing," he revealed. "The initial plan was for me to command the first circumlunar mission, together with Oleg Makarov, in June or July 1967. We then expected to be able to accomplish the first moon landing - ahead of the Americans - in September 1968." The race was getting tight.

The number of cosmonauts assigned to the two lunar programs was often fluid, with cosmonauts taken from the group to train for Earth-orbiting Soyuz missions as they were needed. From early 1967 regulars training alongside Leonov and Makarov in the L1 lunar group included veteran cosmonauts Valery Bykovsky, Georgi Grechko, Pavel Popovich, Vitaly Sevastyanov, Nikolai Rukavishnikov, and Boris Volynov. As the L1 ground simulator wouldn't be ready for use by the cosmonauts at Star City until early 1968, they often found themselves spending hours in the Moscow planetarium familiarizing themselves with the constellations. Some of the lesser-known faces amongst them were even sent on clandestine trips to Somalia to see the stars of the southern sky for real.

Oleg Makarov was a civilian engineer, a person General Kamanin and the military cosmonauts had earlier resented. He had been one of the highly-qualified engineers Vasily Mishin had selected for space training from his own design bureau, which at the time had thrust the Chief Designer into bitter disagreement with Kamanin. Ironically, Makarov was now slated to spend days sitting beside Leonov inside a cramped L1. Buoyed by the apparent success of their Voskhod flights, the Soviets had considered their manned capsules safe and airtight; safe enough in fact that they could launch crews without their bulky protective spacesuits. As a result, cosmonauts on L1 missions were expected to undertake their assignment dressed simply in lightweight woolen tracksuits. Only the N1-L3 crew would take along spacesuits, designed for the transfer spacewalks in lunar orbit and the all-important moonwalk.

Although the loss of the three Apollo 1 astronauts in a spacecraft fire during a countdown test in January 1967 was a terrible tragedy, the Soviets must have secretly seen it as offering some much-needed time to catch up with the Americans. A month before their own Soyuz 1 disaster leveled the playing field once again they were ready with the first test flight of their moon ship.

The L1 version of the UR-500 (now know as the 'Proton' booster) made for an impressive sight as it stood some 200 feet tall on its launch pad. The mighty booster consisted of a new third stage and a Block-D fourth stage designed to rocket the L1 to lunar velocity. Although this first four-stage configuration worked perfectly when Cosmos 146 (as it was misleadingly designated) lifted-off on 10 March 1967 in a first test of the L1 hardware booster, many within TsKBEM still wanted manned missions to be dual launched. This concept would see the L1 placed in orbit unmanned, with the crew launched separately on a safer Soyuz rocket before transferring to the L1/Block-D vehicle during a spacewalk. This plan would have been very uncomfortable for the two spacesuited cosmonauts, as the L1 only had one cramped cabin. By contrast the Soyuz, with its spacious frontal orbital module, had almost three times the living space.

General Kamanin missed this first L1 launch, so he made sure he took the L1 cosmonaut group to see the next lift-off from Baikonur, on 8 April 1967. For most of them it was the first time they had seen a Proton rocket up close. Their amazement was only marginally tainted when the L1/Block-D mock-up payload became stranded in Earth orbit after a malfunction and given the nondescript cover designation of Cosmos 154. More serious was the tragic death two weeks later of veteran cosmonaut Vladimir Komarov on Soyuz 1. This had major repercussions for the L1 as they were essentially the same spacecraft. Much soul-searching must have been done because the back-up parachute on the L1 had been removed in order to save precious weight.

Any hopes for a manned flyby of the moon to celebrate the October Revolution in 1967 had now vanished, but while the Americans could only guess at their space rival's intentions, the

Soviets had the advantage of full access to American documentation. One surprising example of their intimacy with the American technology came when L1 cosmonaut Pavel Belyayev paid a courtesy call on the American pavilion at the Paris Air Show in May 1967. He jumped into the Apollo mock-up on display and amazed his hosts by seeming to know his way around the interior and displaying an unexpected familiarity with its systems. At the same show Belyayev also let slip to Gemini 10 astronaut Michael Collins that he expected his next flight to be to the moon and that he was practicing unpowered helicopter landings - something only useful for aspiring lunar lander pilots who wanted to see what a steep moon landing descent was like. We now know several cosmonauts completed these dangerous flights using a helicopter fitted with the same type of digital computer that would have been installed on the Soviet lunar lander.

As if to confirm the stalling of both moon efforts, during the next Proton launch on 28 September 1967 one of the first stage engines failed to ignite and the rocket flew off-course for about one minute before the emergency escape system dragged the L1 to safety. Although the capsule landed some forty miles downrange, the recovery team (which included Alexei Leonov) was hindered by fumes from the crushed rocket tanks nearby. The 'Devil's Venom' was living up to its toxic reputation. The recovery team had to drive around the flat terrain of the steppe searching for some high ground in a frantic attempt to escape a creeping cloud of toxic gas. The cosmonauts must have wondered if they'd face the same problem if they ever had to survive a malfunctioning launch.

On 22 November 1967 the next unmanned L1 launch saw the second stage malfunction, but this time the capsule was heavily damaged when it plowed into the ground at high speed 177 miles downrange.

In a bid to inject some urgency into the project during 1968, Mishin planned to launch one unmanned L1 a month starting in March with the aim of sending a cosmonaut to the moon by the end of the year. The first launch on 2 March saw the ship, designated Zond 4, successfully boosted to a lunar distance by its Block-D stage. It had been deliberately launched in the opposite direction to avoid any complications caused by actually going anywhere near the moon, and its true purpose as a test of a spacecraft intended for transporting cosmonauts was disguised by recording it as one of a series of scientific biosatellites. Some of the L1 crews who had been at Baikonur for the launch of Zond 4 then flew to the Yevpatoriya tracking centre for communications tests with the craft.

As the capsule looped out to lunar distance, the cosmonauts decided to toast the health of the craft with champagne. They also took time to toast Yuri Gagarin's thirty-fourth birthday, occurring just a few days later. Vitaly Sevastyanov remembered that it was only "a tiny glass of champagne, because we had to do some work," but even these modest celebrations were premature. What initially seemed like success over lunch turned to failure at the last possible moment when Dmitri Ustinov of the Defense Ministry of the USSR ordered the capsule's destruction during re-entry after a sensor indicated it was off-course and could end up falling into the ocean. A sophisticated 'double skip' re-entry into Earth's atmosphere on the way back from the moon was essential to guarantee a landing in the Soviet Union. Unfortunately the Soviet military had balked at the costs involved in any such ocean recovery operation when they realized it would involve over 16,000 navy personnel, and there were no ships available to recover the Zond craft if it had been allowed to splash down. Traditional Soviet paranoia now dictated that it be blown up in case it fell into the eager hands of a foreign power. This was especially frustrating for the cosmonauts and its designers as they knew any crew aboard would have survived the 20-g ballistic re-entry.

If the destructive loss of the Zond spacecraft wasn't depressing enough, the cosmonaut corps was deeply shocked only weeks later when Yuri Gagarin was killed in a fiery jet crash. Although the lunar pairings had back-up pilots, able to step in at the last minute to replace the commander, no one had seriously considered this option until Gagarin died. As a result, the Kremlin grew increasingly wary about risking any more heroic household names, especially the world's renowned first spacewalker. Leonov's reserve Anatoli Kuklin suddenly found himself being prepared for a manned Zond mission during 1968.

"On the 27 March Yuri Gagarin died. Our chief started to think things over," Kuklin revealed in a 2001 interview with researcher Rex Hall. "The first cosmonaut had died, and now, there was the first cosmonaut to walk in space, Alexei Leonov ... [our chief] said 'Let's assign Kuklin,' so I trained in March and April. One flight-engineer and two commanders: poor Makarov underwent two training cycles."

Luckily for Leonov, continued problems with both the Proton and the L1 craft meant that any decision to continue was held over in the corridors of power. That June a list of cosmonauts selected for the N1-L3 landing missions was finalized. Along with cosmonauts already training for L1 flights, pilots selected included Viktor Gorbatko, Yevgeni Khrunov, Viktor Patsayev, Anatoli Voronov, and Alexei Yeliseyev. The lessons of the Zond 4 fiasco were well learned: when the next unmanned L1 lifted-off on 23 April 1968, ten Soviet Navy ships were dispatched to the Pacific Ocean. However a troublesome second stage on this unmanned test flight meant that the payload failed to reach Earth orbit. On a more positive note the escape system worked perfectly once again and the capsule touched down safely on the steppe, not far from Baikonur.

The next Proton launch attempt, set for late July 1968, almost caused a catastrophe on the scale of the R-16 explosion back in 1960. With more than 150 technicians working on the launch pad, the rocket's Block-D liquid oxygen tank burst after it was over-pressurized and the top part of the rocket containing the spacecraft began to topple over. Momentarily it looked as if the whole booster might explode around the pad workers, but amazingly the L1 escape tower became entangled in the pad gantry, preventing it from falling onto the fully-fuelled booster below. Although it would take several weeks to dismantle and remove the twisted rocket it could have had a far worse outcome.

By the time Zond 5 lifted off successfully from Baikonur on 15 September 1968, Mishin's schedule was in dire trouble. To that time the unmanned tests had mostly proven to be failures; so much so that the Soviet Navy decided once again to cut back on its ocean recovery fleet in order to save money, declaring that the odds against the unmanned L1 actually making it back to Earth looked very slim.

But the craft - which this time included a 'crew' of two tortoises – successfully achieved Earth orbit before being rocketed towards the moon by the Block-D stage. It was the first time one of the unmanned L1s had progressed to this stage and cosmonauts such as Bykovsky and Popovich were on hand once again at the tracking station to test voice communications with the spacecraft. These transmissions, with Popovich playfully pretending to be in communication with an on-board cosmonaut, caused great confusion in the West when they were detected. Intercepting these communications from deep space did at least confirm one worrying, growing suspicion - that the Soviets were preparing for their first manned circumlunar flight.

The Soviet Navy's decision to scale back its recovery fleet would backfire badly on them when Zond 5's re-entry guidance system malfunctioned on the return journey and it had to make a ballistic descent into the atmosphere. When it splashed down in the Indian Ocean at night, the nearest Soviet ship was some sixty-five miles away. As a result it was several hours before the bobbing capsule could be retrieved from the ocean. Embarrassingly, this recovery all took place in front of a group of U.S. Navy ships which had also been alerted and had arrived at the scene. Apart from giving the Americans a good look, this mission was the first triumph for the L1 program, as a spacecraft had not only travelled to the moon and back, but it had also captured good-quality, close-up images of the moon. Furthermore, it had proved that any technical problems could be overcome to bring the ship home safely. Adding to the elation of the mission scientists was the fact that when the spacecraft was opened up, the two tortoises were recovered alive, having survived their lunar loop.

Soon after, when Nikolai Kamanin heard rumors that NASA was examining an audacious plan to send only its second manned Apollo spacecraft to the moon, he was shocked at its sheer adventurism. "One can understand the Americans," he recorded in his diaries. "Their Apollo cannot fly around the moon unmanned. In order to test their ships the Americans are forced to risk the lives of the crews, but we, having ships like Zond 6, can test them without such risk."

However even Kamanin must have realized it was the Soviets' own insistence on conducting several unmanned test flights before putting men aboard the L1 that had cost them dearly in the race to the moon. If only they had shown less caution and allowed cosmonauts aboard earlier, they might very well have been the first to send men around the moon in 1968. They certainly had that capability, as the Zond 5 mission had proved. The only forlorn hope they could now cling to was that Apollo 8 might somehow fail in its attempt to orbit the moon, leaving the way clear for a Soviet manned mission in early 1969.

Unfortunately, there was another stumble in the Soviet plans. Immediately following the launch of Zond 6 on 10 November 1968, the craft's high-gain antenna failed to deploy as the Block-D blasted it towards the moon. Although this didn't prevent Leonov from conducting communications tests with the lunar craft at Yevpatoriya, it made life for the engineers on the ground particularly difficult.

On 12 November NASA finally confirmed that Apollo 8 would head for the moon in December, so the race was essentially still on as Zond 6 passed around the moon just two days later, recording impressive images of the lunar landscape below.

Kamanin openly discussed the situation with the anxious cosmonauts, telling them that although they were prepared for their mission, a flight-ready version of the L1 wouldn't be available until January 1969. In an attempt to boost morale he asked them to provide a new name for the manned L1. Normally, the superstitious cosmonauts wouldn't have been involved in this process, believing it could be some sort of bad luck omen. As Oleg Makarov explained, "We had a tradition that the names are chosen only a few days before the launching. Before that, privately, everyone used the factory codes L1 and L3." Now for the first time more romantic names were being openly discussed. Suggestions included 'Rodina' (Motherland), 'Ural' and 'Akademik Korolev'.

Even as names for the first ship were being debated, things began to go wrong aboard Zond 6 when the hydrogen peroxide supply for the control thrusters froze. This situation, if not resolved, would make the spacecraft uncontrollable during re-entry. In an effort to thaw out the hydrogen peroxide, controllers transmitted instructions that turned the craft's tanks towards the sun. However this innovative measure also exposed a faulty rubber seal around the main hatch to the heat, causing the air to begin to leak out of the cabin. Although the cured thrusters initiated the sophisticated double-skip re-entry needed to land inside Soviet territory, an electrical short-circuit during the landing caused the parachutes to be automatically cut free while it was still several miles above the ground.

The free falling spacecraft slammed into the ground at high speed. Fortunately the retro-rockets had also fired early, so unlike Soyuz 1, the Zond 6 capsule did not explode on impact, which meant that some of the scientific experiments were safely recovered. Nevertheless, it was abundantly clear and deeply troublesome that had there been cosmonauts aboard, they would have perished. Ironically, photographic film recovered from the wreckage was developed and shown around the world to give the impression that the mission had been a complete success. By the end of 1968, NASA was ready to send Apollo 8 to the moon. It was a devastating time for Alexei Leonov. "I was in Moscow, busy working on the L1 circumlunar program, when news came through that the Americans had sent a manned spacecraft into orbit around the Moon," he recalled. "I suddenly had the feeling that everything was slipping through my fingers. I could see my dreams going up in smoke."

Others who had flown to Baikonur to take part in an upcoming Soyuz mission couldn't help but stare longingly at the moon as three American astronauts slipped into orbit around our nearest celestial neighbor. "We saw the bright crescent of the moon and for a minute we were all silent," Kamanin wrote in his diary on 24 December. "We were filled with contradictory feelings: it hurt that not our guys were first around the moon but [nevertheless] we all admired the courage of the American astronauts and silently wished them success."

At an official meeting on 30 December to suggest any immediate response to Apollo 8, the topic of the L1 wasn't even raised by those in the room. Many still dreamt of a Soviet moon landing mission in the 1970-71 period, believing that the Americans might not enjoy a successful landing on their first attempt. But even at a time when a circumlunar fly-by was seen to be irrelevant, scarce resources continued to be wasted on unmanned Zond flights, as spacecraft designer Igor Afanasyev discussed in a 1991 article.

"Immediately stopping a flywheel once it has been started is virtually impossible," he reasoned. "A program that had produced very satisfactory results couldn't be cancelled. Besides, the spacecraft were built, the launch vehicles were waiting. The schedule of flights had to be observed."

With plans for a manned L1 flight now abandoned, Kamanin and the L1 cosmonauts had to be content with watching the next Zond spacecraft taking off from Baikonur on 20 January 1969. This flight used the refurbished descent capsule salvaged from the previous (23 April 1968) launch failure. It was with some irony that this capsule had to be pulled free, once again, by the escape system from an errant Proton booster eight-and-a-half minutes into the flight. This time it landed in a valley in Mongolia and was recovered three hours later.

General Kamanin confided to his secret diary his growing frustration that the cosmonauts were being denied their chance to fly around the moon. "I dreamed of seeing the day when I would fly to Moscow with our guys after their return from the moon," he wrote. "These were completely realistic dreams but major mistakes by the leaders of our space program and excessive automation of spaceships let to the Americans jumping ahead and flying around the Moon on Apollo 8." In early 1969 General Kamanin also expressed surprising rage at what he saw as the downplaying of historic American achievements in the Soviet press. A prime example came, he wrote, when Apollo 10 was unfavorably compared to unmanned Soviet probes sent to the planet Venus. "While little is written here about this outstanding mission, the successes of Venera are being drummed into us," he observed. "Although even to Boy Scouts it's clear that the importance of the Apollo flight is ten times greater than the flights of all our Veneras."

To have any rapidly-dissipating hope of matching Apollo, the N1 had to be launched in early 1969, but many of its designers still didn't believe it was ready. Only enormous pressure from Mishin and his supporters resulted in a February launch date being set. Movie footage of the roll-out of the booster on its huge horizontal transporter, mounted on twin railway lines, shows onlookers with nervous smiles on their faces. Perhaps they were simply relieved the day had finally come to see if it could actually fly.

This first N1 to fly was codenamed '3L' and its three painted green-and-white conical stages caused it to resemble a military missile as it took off into the early afternoon sky on 21 February, carrying a modified L1 and Block-D stage under a 140-foot-long white fairing. An explosion in the first stage engine compartment at sixty-nine seconds into the flight resulted in the loss of the whole vehicle. As on numerous Proton failures previously the escape system worked perfectly, dragging the spacecraft mock-up safely away as the N1 debris arced over before impacting the ground thirty-one miles downrange. Amazingly only two of its thirty first-stage engines had failed to work on lift-off.

After years of waiting for this very moment, the CIA managed to miss the entire show as this short flight literally flew under the radar of American monitoring stations in Turkey. It would only be later, when old satellite photographs were closely examined, that the large crater made by this crash was discovered.

Optimistically, in June 1969 Kamanin and Mishin selected a final group of eight cosmonauts for moon landing missions. They were Bykovsky, Khrunov, Leonov, Makarov, Patsayev, Rukavishnikov, Voronov, and Yeliseyev.

After studying the first N1 failure it was decided to carry out extensive modifications to the 4L booster, so it was replaced with the newer 5L for the second launch. This carried another L1 mock-up as it lit up the night sky on the evening of 3 July 1969. Then, less than ten seconds into the flight, a shard of metal entered the oxygen pump of one of the engines, creating an explosive chain reaction which forced the KORD computer to shut down all the remaining engines. When this occurred the booster was only 650 feet above the launch pad. Slowly, inevitably, gravity took hold and the giant booster slumped back to the ground before a massive detonation took place which knocked stunned observers a dozen miles away off their feet. The flight had only lasted twenty-three seconds but when it was all over, so was any last hope of a Soviet cosmonaut on the moon in the early 1970s.

That accident effectively destroyed the main launch pad, in addition to damaging the assembly building and an N1 mock-up sitting on the back-up pad nearby. For some of the cosmonauts present - including Leonov, Makarov and Khrunov, Rukavishnikov and Khrunov - it was painfully obvious to them that they would probably never walk on the moon. Years later Khrunov admitted that he'd wept after the accident as he realized it was all over.

At that same time, less than two weeks before the Americans would attempt their first manned moon landing, astronaut Frank Borman was in the Soviet Union on a nine-day semi-official tour organized by the Institute of Soviet-American Relations. His scheduled visit to the cosmonaut training centre near Moscow couldn't have been more poorly timed as it was due to take place less than thirty-six hours after the launch of this latest N1. General Kamanin, who had refused to attend the launch, was on hand to greet Borman personally, but cosmonauts who had witnessed the explosion kept what they had just seen at Baikonur to themselves as they sat through Borman's presentation of his Apollo 8 'home movies.'

Borman must have wondered about the sea of gloomy faces surrounding him. Alexei Leonov arrived later and managed to catch up with Borman at a dinner in Moscow's plush Metropole restaurant. He noted that it was packed full of uniformed military personnel, there to see the famous American. "Everyone wanted to stand near him. To touch him," Leonov remembered. "When I eventually met him I congratulated him on his mission. I of all people, I said, knew how hard it must have been. I did not tell him I, too, had been training for a circumlunar mission. But I felt as if he knew [and] we discussed good locations for lunar landing."

Ironically, the favored landing site for the first Soviet mission would have been the Ocean of Storms, the eventual landing site selected for Apollo 12, while the Sea of Tranquility was actually their third choice. These had been independently selected by scientists at the Vernadsky Institute in 1968 using data gathered from several early Luna probes.

In many ways the N1-L3 landing would have been very similar to that of Apollo, consisting of the LOK (Luniy Orbitalny Korabl) mother ship and an LK (Luniy Korabl) lunar lander. The LOK looked to the untrained eye like a longer version of the standard Soyuz without its solar panels, but it was almost totally different internally. The mother ship was designed by a team headed by Yuri Semenyov (later to take over as the head of the space program following the death of Glushko) and included a roomy orbital module. This was needed to store the bulky moon suits and also act as an airlock for a cosmonaut transferring to the lunar lander during a spacewalk. If this had taken place, it would certainly have been one of the most awe-inspiring moments in human spaceflight as it would have taken place in lunar orbit only thirty-seven miles above the surface. Only the knowledge that a risky 'seat of the pants' descent was next would have focused the mind of the spacewalking cosmonaut. The Soviet lunar lander was about as tall as its Apollo counterpart but was only about a third of its mass, as most of its powered descent was performed by the attached Block-D stage. This gave the Soviet moon lander a lean, top-heavy appearance, but with designers conscious of the safety features needed to protect its solo pilot it incorporated a few extras when compared to NASA's LM. The LK's final descent itself would have been very short. Once the Block-D had separated just above the lunar surface, after firing to brake the lander out of its lunar orbit, the cosmonaut would only have somewhere between twenty-five seconds and one minute to find a safe landing site for touchdown. Up to that point most of the landing was highly automated, using the same computer tested by the pilots on helicopter training flights.

In reality, however, it has been calculated that the cosmonaut only had about three seconds to make the decision about where to set his craft down on the lunar surface. Thankfully he could land on up to a twenty-degree slope, as the LK was fitted with four upward-facing solid propellant thrusters which fired on touchdown to firmly plant the lunar lander legs onto the surface of the moon, keeping the lander in an upright position.

As an added safety feature the designers at TsKBEM had also fitted the LK with a single descent/ascent engine (Block-E), so the cosmonaut didn't have to waste valuable seconds starting a second engine if he had to abort. All he basically had to do was throttle up once again to make it back to the LOK waiting in lunar orbit. As well, the Soviet lander was fitted with a second back-up ascent engine in the event the main engine failed to ignite.

Had it taken place, the first Soviet moonwalk would have lasted about ninety minutes. As on Apollo 11 a television camera fixed atop the exit hatch of the LK lander would have recorded the historic moment as the lone cosmonaut stepped on the surface, planted the red flag of the USSR, took a congratulatory call from Leonid Brezhnev, installed scientific instruments and gathered some lunar rocks for the return journey back to Earth.

As it happened, the only piece of moon landing hardware ready to fly by 1969 was the lunar spacesuit. Designated Kretchet after a large arctic bird of prey, this was a radical semi-rigid torso design which allowed the cosmonaut to step into the suit through the backpack and shut it like a door behind him. It was an elegant concept that overcame the lack of room inside the lander. Before his moonwalk the cosmonaut would attach a bizarre 'hula-hoop' tube around his waist, preventing him from becoming trapped on his back like a stranded turtle if he fell over on the moon's surface. The second member of the L3 crew, who would remain in lunar orbit, had a lighter spacesuit called Orlan, which he could use in an emergency to help the lunar surface cosmonaut if he got into trouble during the spacewalk between the LOK and the LK.

By late 1968 the cosmonauts were training under lunar conditions with the Kretchet suit on one-sixth gravity simulations aboard Tupolev 104 aircraft. Some of them had even donned the spacesuit at the Kamchatka Peninsula in the Russian Far East in order to test its effectiveness whilst clambering around the area's rugged volcanic slopes. The Kretchet could have been used on the moon in 1969, if only there had been a spaceship for the cosmonauts to travel on.

Realizing that a space station was the most likely follow-on project, from 1970 the spacesuit's manufacturer Zvezda worked on modifying the design for use in Earth orbit. It would become the basis of the EVA suit successfully used since 1977. Recently it has become the standard Russian spacesuit aboard the International

Space Station and has even been used by American astronauts from time to time. Many consider it one of the best-ever spacesuit designs.

As the world watched the Apollo 11 saga unfold in July 1969, the often forgotten Soviet response, Luna 15, was staged as a last-gasp effort to land, scoop up and return some lunar soil to Earth ahead of the manned landing. Although it was an unmanned probe, a successful mission by Luna 15 would - to a degree - have upstaged the historic American effort. To achieve this, its NPO Lavochkin designers had taken the landing stage of a Lunokhod rover and hastily designed a drill and an Earth-return vehicle fitted with a basketball-sized recovery capsule to carry back a small soil sample. Launched on 13 July, just three days before Apollo 11, celestial mechanics dictated that their sample of moondust would arrive back on Earth a few hours after the Apollo astronauts had splashed down. If successful, it was felt, the flight of Luna 15 might draw some attention from the simultaneous American effort to place Armstrong and Aldrin on the moon by demonstrating the effectiveness of sending robotic machines instead of risking human lives.

Space historian Brian Harvey even discovered details of the bizarre propaganda show waiting back on Earth if it returned safely: "Once the launch was successful, preparations were put in train for a triumphant parade through Moscow, probably for 26th or 27 July. An armored car, covered in the Soviet flag and bedecked with flowers, would bring the rock samples from Vnukuvo Airport into Moscow, past the west gate of the Kremlin and on to the Vernadsky Institute where they would be displayed to a frenzy of the world's press before being brought inside for analysis."

The launch of Luna 15 caused immediate anxiety in the United States once its objectives had been revealed, and assurances were sought from the Soviet Union that it would not interfere with or even endanger the manned mission. It was with some relief that these assurances were issued by the Kremlin. But any plans and propaganda strategies went sadly awry when problems were encountered with Luna 15. It had entered into an elliptical orbit around the moon which brought it to within ten miles of the lunar surface at its lowest point, at one stage practically grazing the area around the Sea of Tranquility. Then, within hours of Armstrong and Aldrin setting foot upon the moon, Luna 15 began its powered descent but slammed into the aptly named Sea of Crises at high speed and was destroyed.

A month after the triumph of Apollo 11, Leonov and Makarov found themselves at mission control acting as operators for the unmanned Zond 7 which had been launched on 8 August 1969. Frustratingly, its living occupants consisted of four tortoises. Although two years late, this turned out to be the first totally successful L1 mission and energized its supporters enough to lobby the Kremlin to fly a manned version around the moon to commemorate the centenary of Lenin's birth in April 1970. The idea was quietly shelved. Deprived of the opportunity to impress the world with anything comparable to Apollo 11, Leonid Brezhnev was reduced to using the unflown hardware as vast props for a piece of political theatre he staged to impress a visiting delegation of Czech Communists touring the Baikonur Cosmodrome that October. They were shown a Lunokhod lunar rover and an L1 before being taken to see the repaired N1 mock-up mounted on the surviving launch pad. As this display of Soviet technology couldn't be spoiled by failure, the twisted remains of the other launch pad destroyed by the July explosion were skillfully hidden using strategically-placed fencing. They would be the last foreigners shown anything related to the Soviet manned moon program for the next two decades.

Around this time Nikolai Kamanin confided once again to his diary that the scarce coverage of Apollo in the Soviet press had spoiled his mood. "The fulfillment of a dream that has captured the imagination of the whole planet for the past ten years has ceased to interest our leaders only because the ones flying to the moon are not we, but the Americans," he wrote. "I cannot join this conspiracy to remain silent about this great achievement of mankind and with all my heart welcome the successes of our American colleagues, although I know the Soviet cosmonauts have a tough time coping with our defeat."

Kamanin's emotional feelings combined with sadness for his own cosmonaut team. Like him, they now realized that their slim chances of going to the moon were effectively blocked by a disinterested Kremlin. As if to confirm this, the cosmonaut corps in training for lunar missions was disbanded and the men transferred to the space station project by May 1970. Leonov was particularly bitter. "I argued very hard that we should continue with our work but the higher powers were adamant," he recalled. "The lunar groups which I had commanded and trained with for three long, hard years were disbanded...There was no use, as the Russian saying goes, rubbing ashes on my forehead - crying over spilled milk. I knew I had to accept the decision. If I did not, there was a risk that I would never fly again."

Any slim chance of a future manned lunar mission all but disappeared on 24 September 1970, when the Luna 16 probe successfully returned some lunar soil to Earth, confirming the Kremlin's proclaimed stance that the Soviet Union's only interest in the moon had been in planning and designing a cheaper unmanned alternative to Apollo. There had never been any Soviet plans, it declared emphatically, to send cosmonauts to the moon.

In that atmosphere the flight of Zond 8 on 20 October 1970 can be seen as a final sad footnote, but at least the project went down fighting. Its biological cargo of tortoises, flies and plants ended their lunar flight with a planned splashdown in the Indian Ocean, demonstrating that the Soviets could have successfully sent men around the moon in 1970, if only the political will had been there. Years later, pictures of what might have been the manned ship emerged from the archives, showing an L1 with the name Zond 9 boldly emblazoned on the side in large red letters. The poor performance of the Proton rocket arguably cost the Soviets the honor of flying the first men around the moon. Despite spending the equivalent of more than two billion dollars on the first nineteen Proton rockets for the lunar project up to February 1970, only six of them could be regarded as totally successful. It was a miserable record that cost them dearly. But in the meantime Vladimir Chelomey, who had been concentrating on a military space station project called Almaz, had the last laugh when this design and the Proton booster were used by the Feoktistov faction within TsKBEM as the basis for the new Salyut space station.

Vasily Mishin, for his part, had been opposed to any Proton-launched space station project from the start, knowing it would deflect attention away from his faltering N1. Following the launch of the first Salyut station on 19 April 1971 he had shown a distinct lack of interest in the station program, which some later saw as a contributing factor to the subsequent Soyuz 11 disaster. The inexperienced, under-trained crew of Soyuz 11, which consisted of former L1 candidate Georgi Dobrovolsky, together with Vladislav Volkov and Viktor Patsayev, had surprisingly replaced Leonov's prime crew in its entirety only three days before the launch due to a possible medical problem with crewmember Valery Kubasov. It was a surprise to many when Mishin took some of his senior engineers with him to Baikonur for the launch of the third N1 when they should really have been on hand at mission control to help support the first Salyut space station crew, who were far less familiar with the station's systems and hardware than the original prime crew.

Mishin had openly demonstrated his first love when he ordered the cosmonauts onboard Salyut 1 to watch the launch of the 6L booster, carrying a LOK capsule, as they passed over Baikonur on 20 June 1971. Perhaps symbolically, the giant booster refused to co-operate and Mishin had to postpone the N1 lift-off. When it finally left the pad on 27 June 1971 it did not have an audience above, but the booster, now painted pure white like the Saturn V, was a spectacular sight as it lit up the night sky on a pillar of flame nearly three times its own length. In a bizarre twist this launch failed because all of the engines worked perfectly, creating a previously unforeseen roll motion which couldn't be counteracted. Eventually rotational dynamic forces on the whole booster grew so strong that the top stages literally sheared off and crashed near the launch pad. The remaining section thundered away in a giant arc, eventually impacting and exploding downrange.

Mishin finally took charge of the Soyuz 11 mission at that point when he flew directly from the N1 failure to the Yevpatoriya tracking centre to supervise the crew's return at the end of their highly successful three-week mission just three days later. Subsequent commentators have linked Mishin's obsessive focus on the falling moon program, when he should have been concentrating on the new space station, as one of the reasons for the safety lapses which killed the Soyuz 11 cosmonauts when their cabin depressurized on re-entry.

After these two disastrous events, the loss of the third N1 and the deaths of the Soyuz 11 crew, Mishin's authority diminished rapidly. With the current N1-L3 project looking increasing irrelevant, he had to start promoting a follow-on 'lunar outpost' project called N1-L3M that utilized two N1 launches and a larger, multi-person lunar lander capable of longer stays on the moon. But even he admitted that none of this could be ready until the late 1970s.

Ironically, Mishin was in the hospital and missed what would turn out to be the last and best N1 launch on 23 November 1972. Less than two weeks before America's final manned lunar mission, stand-in launch director Boris Chertok must have felt that success was at last possible with the improved 7L. This latest rocket was fitted with new roll control motors to prevent a repeat of the previous failure, and over thirteen thousand sensors to monitor the progress of the flight. After lift-off the N1 climbed smoothly for 107 seconds before an unexpected vibration in the propellant lines for engine No.4 caused them to rupture, and the ascending booster disintegrated in a massive fireball. The frustration of everyone involved must have been intense as the explosion took place only seven seconds before the scheduled first stage cut-off. As the escape tower once again pulled its payload to safely, the contrast with the Americans couldn't have been more evident. As NASA was winding down Apollo as a technological and political triumph, the Soviets had not managed to fly their own moon rocket successfully even once.

During a trip to Moscow by moonwalker Dave Scott in 1973, in preparation for the Apollo-Soyuz Test Project (ASTP), Leonov couldn't contain himself any longer and over a quiet drink at his apartment told the astonished astronaut of his own lunar training. Although NASA's senior management had been given CIA briefings on the subject, this had never filtered down to the astronauts and much of what Leonov confided to Scott was news to him, as the astronaut later wrote in his *Two Sides of the Moon*. "It was fascinating to learn that the Russians had been that far along the path towards a lunar landing and to learn that Alexei was their key man," he revealed. "Our mindset at the time was that the Russians did not tell anyone anything, so the openness with which Alexei and I talked that night was, to me, quite fascinating."

The improved 8L and 9L boosters, using the modernized NK-33 Kuznetsov engine, were being readied in 1974. But just as the 8L booster, the fifth N1, was being prepared for launch sometime in August 1974, Mishin's enemies finally pounced, as described by Nikolai Kamanin. "Many mistakes were made in the L3 project; they were visible even before the appearance of Apollo and after the first successful flights of it later it became clear to everyone that our lunar ships cannot compete with it," he wrote in a diary entry. "The sad story of the N1 and L3 is not finished yet: Mishin and his high patrons are trying to 'cure' a bad rocket and a bad ship and at the same time we are continuing to lag ever further behind the USA."

Kamanin and many others wanted to kill off the N1 project before it successfully placed something into orbit. They knew that once that happened they would be stuck with a rocket they all knew was flawed. With the tragic deaths of the returning Soyuz 11 crew on his hands and the failure to match the new American space station Skylab, Mishin's enemies now had enough ammunition to finally get him sacked by the Kremlin.

"There was no political advantage in continuing to go to the moon because the Americans had already been there six times," rationalized space designer Boris Chertok. "If we had stepped on the Moon even once after that, nobody would have been particularly happy. So it would have been necessary to plan a new program - build a real lunar outpost - and that would have been very costly. The political leaders then in power were not prepared to go along with that."

By this time, most problems with the four-stage Proton had been overcome. Three LK lunar landers launched by Soyuz boosters had been tested successfully in Earth orbit during 1970-1971. If only the Soviets had begun building all these spacecraft a few years earlier, then the Americans might have had a real challenge when it came to Kennedy's lunar goal of a safe landing and return by the end of the decade. Now, suddenly, there was the decision to suspend N1 flights, followed by the order to destroy the two remaining flight-worthy N1 rockets.

One authoritative estimate quotes a figure of four billion rubles for the construction of ten N1 rockets, so each one lost would have cost an unsustainable 400 million rubles. Considering the fact that the Gross Domestic Product of the USSR at that time was between one quarter and one-half that of the U.S. economy, this was a serious expense the Soviet Union could ill afford. That aside, terminating the N1 program was a massive blow to those involved. "Six rockets, two of them fully assembled, were dismantled," Vasily Mishin commented with obvious bitterness. "The people who devoted the best years of their lives to them cried." He would later say that he understood everything - except Valentin Glushko taking over as the new Chief Designer.

Although Glushko is often portrayed as the villain of the piece the race to the moon was over and, he more than anyone, realized that a new booster was needed for a new era of space shuttles and manned missions to Mars.

One subsequent order of Glushko that can be viewed as totally vindictive was his instruction to Nikolai Kuznetsov to destroy all the remaining N1 engines. There was little sense in doing this; by that time the NK-33 had developed into a superb engine. Engineers had even managed to run one continuously on a test stand for an astonishing five-and-a-half hours. Perhaps Glushko realized he had a potential rival in the making.

Not surprisingly, Kuznetsov was outraged by this order, which he decided to ignore. He then covertly hid over one hundred of his beloved NK-33s in a disused section of his factory under large 'Nuclear Danger' warning signs. That pride in his engines would turn out to be a life-line for the cash-strapped factory in the 1990s, when American engineers from Aerojet declared it one of the best rocket engines ever and purchased dozens of them for a million dollars apiece. Ironically these NK-33 engines were fitted to the latest version of the Atlas launch vehicle - the same American former ICBM used to launch John Glenn into orbit at the height of the Cold War back in 1962.

Soon after achieving the job of Chief Designer, Glushko merged TsKBEM with Energomash to form one giant new organization that became known as NPO Energia, which he kept under tight control. This enabled him to start again from scratch on a super booster design called Energia. Using high-tech liquid hydrogen/liquid oxygen propellants like the Saturn V, the booster made its debut flight in 1987 before successfully launching the Soviet space shuttle *Buran* on a fully-automated, two-orbit mission in November 1988. This superb rocket - which flew less times than the N1 - sadly became one of the first casualties when the Russian space budget collapsed following the demise of the Soviet Union.

Although the twentieth anniversary of the Apollo 11 landing is often regarded as the main spur for an increasingly open Soviet society to finally reveal its past lunar ambitions, it is no coincidence that the controlling influence of Glushko had just ended with his death in January 1989. Others could finally challenge his "official history" of the late 1960s - something he had easily managed as editor of the official *Soviet Encyclopedia of Cosmonautics*.

At first many were still reticent to reveal details about plans to send cosmonauts to the moon ahead of the Americans - a sensitive blank page in space history. Then, in the same month in which Glushko passed away after a prolonged illness, Vasily Lazarev became the first cosmonaut to publicly confirm to a lecture audience the existence of a manned lunar program. Lazarev, who had flown twice with Oleg Makarov, may have been the first to reveal the existence of this colossal Soviet program, but he still refused to divulge more details because it was "a matter of the past."

The first solid facts came in a book about cosmonaut Valery Bykovsky published that summer, which told the story of his training for a manned Zond flight in the 1960s. Once the secret was out, Mishin used a series of newspaper interviews as an apologia for his own failures. "We came just one step from success but we were not allowed to take that step," he told *Pravda* newspaper in October 1989. "The Americans had already invested $25 billion in their program and eventually reached the moon. We invested ten times less and had to fight for every million rubles."

In December 1989 a visiting group of professors from the Massachusetts Institute of Technology (MIT) were shown a Soviet lunar lander during a tour of the Moscow Aviation Institute museum, which had become Vasily Mishin's second home since his forced retirement in 1974. The surprised Americans took some pictures and passed them onto the *New York Times* when they got home. They were published on the front page under the dramatic headline "Russians Finally Admit They Lost Race to Moon."

The moon project might have started out as a dream but it soon turned into a nightmare that not only cost billions of rubles, but failed to give anything back in return to either the Kremlin or the Soviet people. Unlike NASA, everyone at the heart of the Soviet space effort had been reluctant to co-operate with other agencies because they were too busy protecting their own interests and trying to win favor with the Kremlin. Any advances they might have gained by Korolev's series of spectacular space firsts in the early 1960s were quickly wasted by in-fighting over precious funds and who should get what for their own space projects.

Although the cosmonauts would have given anything to try and make a landing on the lunar surface, their superiors eventually believed the odds were stacked against them and cut their losses after the demoralizing success of the rival Apollo program. On the other hand it is a tribute to their skill that they were still within sight of NASA when it came to flying cosmonauts around the moon with the L1.

Ultimately it was the Soviet people who missed out. Denied the truth for over twenty years about how their nation's engineers and cosmonauts had planned to fly to the moon ahead of the Americans, they were also prevented from watching one of mankind's most defining moments as it took place in July 1969.

As Leonov was to later lament, "Not showing live coverage of the Apollo 11 moon landing was a most stupid and short-sighted political decision, stemming from pride and envy. The Soviet Union had been working for so long that there were those who felt they could not show another achieving our goal. But our country robbed its own citizens by allowing political considerations to prevail over genuine human happiness at such events."

Boris Chertok was part of that small group of military men, scientists and cosmonauts who had secretly gathered in the Komsomolsky Avenue facility to watch live pictures of Armstrong and Aldrin walking on the moon. Although only able to witness one of the most monumental events in human history because their small television monitors were hooked up to a bootleg cable system coming in from Europe, he still shook his head in wonder as he watched events unfolding on the Sea of Tranquility. "We were delighted as engineers as they had done a wonderful job," he would state many years later. "But on the other hand we felt disappointment. Why them and not us? It was bitter."

"It was not a fair race," was Vasily Mishin's final conclusion. "First of all, America was richer than we were, especially then, and Russia was weakened by the fight against German fascism and weakened by the cost of the arms race. As soon as America began the moon race, we understood we could not win."

Although Mishin presided over the Soviet Union's failed efforts to beat America to the moon, it is his unfortunate legacy to be widely perceived as a failure after the stunning achievements of his predecessor. Unrepentant and disillusioned to the end of his eighty-four years, Vasily Mishin, perhaps unfairly regarded as the man who lost the moon race for the Soviet Union, passed away in Moscow on October 10th 2001.

Further reading:

Soviet & Russian Lunar Exploration, Brian Harvey, Springer-Praxis 2007.

The Kamanin Diaries 1960-63, Bart Hendrickx, Journal of the British Interplanetary Society Vol. 50 No. 7. *Diaries 1964-1966*, JBIS Vol. 51 No. 11. *Diaries 1967-168*, JBIS Vol. 53 No. 11/12. *Diaries 1969-1971*, JBIS Vol. 55 No. 9/10.

Two Sides of the Moon, Alexei Leonov & David Scott, Simon & Schuster 2004.

Sputnik and the Soviet Space Challenge & The Soviet Space Race with Apollo, Asif Siddiqi, University of Florida Press 2003.

Sleuthing the Space Sleuths

During the Cold War a network of amateur space watchers known as the 'space sleuths' was at the forefront of discovering the secrets of the Soviet space programme. After the Moscow archives opened their work became a footnote in the history books but now their efforts are being remembered once again.

Much of the renewed interest in the work of the space sleuths was caused by the death of Rex Hall in May 2010. Not only was he one of the west's top experts on the cosmonaut team, he was also the respected chairman of the BIS (British Interplanetary Society) 'Soviet Forum' so his passing seemed like the end of an era. In an attempt to preserve some of his best 'war stories', I set about contacting many of his friends and colleagues for their memories of space sleuthing. Although these were published as a nine-page feature in the June 2011 issue of *Spaceflight*, I felt there was still much to be told and pitched the idea for a full book on the subject to a publisher. Soon I found myself editing *Cold War Space Sleuths* (Springer-Praxis 2013).

During the 1960s anyone with more than a passing interest in Soviet spaceflight found a ready market for their theories in the press, as it was only too willing to publish anything that spiced-up the dull basics coming from *Pravda* and *TASS*. Although western intelligence agencies knew fairly accurately what was really going on behind the Iron Curtain, they rarely revealed the truth for fear of exposing details of their spy networks.

Firstly, who are these 'space sleuths'? Surprisingly the term is not new and was in use as early as 1963 to describe anyone who studied Soviet spaceflight on an unofficial basis. Although they are often referred to as 'amateurs', it is important to remember that they were often experts in other fields and brought a leftfield approach to the study of Soviet spaceflight. This lateral thinking often resulted in correct answers being found just as quickly as the 'professionals'.

By the 1970s their successes were increasingly due to the influence of English school teacher Geoffrey Perry, whose radio tracking of newly launched Soviet satellites provided a constant stream of 'scoops' to the media. He is best remembered for his 1966 'discovery' of a new launch site at Plesetsk, with the growing network of amateur satellite trackers he inspired called 'Kettering Group' in his honour.

Surprisingly we now know that Geoff Perry and others were being subtly guided in the right direction by Dr. Charles S. Sheldon II of the US Library of Congress Research Service. Although Sheldon had top-level security clearance himself, he was hampered by the fact that he could only use information in his own reports that had already been printed in the open media. In an attempt to get around this, he used his own informal contacts with the sleuths to gently point them in the right direction – confident that they were likely to discover what he already knew. Once they published their findings, Sheldon was safe to include them in his own reports without compromising his security vows!

Although the BIS has 'British' in its title, it has always had a truly international membership and has a reputation for non-biased coverage of all space programmes. For this reason it is no surprise that the society became an unofficial base for the space sleuths.

The society's own dealings with the Soviet space establishment go back to the early days of the space race, with Yuri Gagarin being presented with a BIS medal in 1961. Ironically, the cosmonaut's official translator during that trip to London was none other than Boris Belitsky - a name familiar to many of the sleuths when he later became Radio Moscow's space correspondent.

Spaceflight magazine's attempts to cover Soviet spaceflight fairly dates back to its first two editors – astronomer Patrick Moore and writer Kenneth Gatland. As long-standing BIS officials, both were able to established close contacts with Soviet scientists. As editor in the 1970s, Gatland was ideally placed to satisfy a growing interest in the pages of the magazine as to why the Soviets had 'lost' the moon race.

Just as Gatland's editorship of *Spaceflight* was coming to an end several London-based members including Rex Hall, Phillip Clark and Anthony Kenden approached the society with the idea of holding a formal event dedicated to the topic of Soviet spaceflight. Although this inaugural meeting only last two-and-a-half hours on a Friday evening in January 1980, the society still didn't know what to make of it. Recently society historian Bob Parkinson admitted in *Interplanetary: A History of the BIS* that it was several years before the society saw it as more than a just 'fan club' meeting. Thankfully the quality of the papers being presented ensured it would eventually grow into a well-respected annual event attracting speakers from all over the world.

My dilemma when compiling *Cold War Space Sleuths* had been to ensure that as many sides of the story as possible were told. Unfortunately many of the pioneering sleuths such as Perry and Sheldon are no longer with us but their roles as mentors would mean their stories would not go unmentioned. Surprisingly even more recent events, such as the origins of the Forum, were in danger of being lost but luckily David Shayler and Phil Clark provided some of the missing details.

Apart from being a permanent record of events, I also hope *Cold War Space Sleuths* contains some of the excitement of discovery and there are plenty of 'Eureka' moments. My favourites include: James Oberg and Charles Vick accidentally discovering how Soviet censors airbrushed failed cosmonauts from official photographs; Vick discovering a mysterious Soviet drawing and finally revealing the true shape of the top secret N1 booster; Phil Clark running around Moscow with a BBC television crew in the late 1980s trying to avoid their KGB handler; and French sleuth Claude Wachtel turning-up in London with an obscure Soviet book that changed the accepted history of the moon race.

Getting that last story was particularly satisfying as I was determined to include a few of the Paris-based sleuths. Unfortunately their work was often neglected, even though they were sometimes ahead of their English language colleagues. Ironically, whilst many sleuths learned Russian to decipher the conversations of orbiting cosmonauts, this dedication didn't extend to learning some French to read the Paris-based journal *Orbite*.

Unfortunately my plans to include some Russian sleuths in the book (yes, they did exist during the Cold War!) fell through but

Bart Hendrickx and Asif Siddiqi were able to tell some of this story in their own chapters. Although the golden era of space sleuthing might now be over, their work is being remembered again with a growing number of memoirs from those directly involved. Sleuthing might only be a footnote to the much larger story but their efforts were still part of the Cold War battle for the truth.

The 'Cold War Space Sleuths' team.

(L-R) Asif Siddiqi, Phil Clark, Bart Hendrickx, Dominic Phelan, David Shayler, Brian Harvey and Bert Vis.

Soviet Forum 2011

The first Chinese/Soviet Forum since the death of its long-standing chairman Rex Hall was not only an ideal occasion for friends and colleagues to remember him but is a sign of the Society's continued commitment to the study of Soviet spaceflight history. Reflecting the truly international make-up of the day over 25 people from Britain, Ireland, Holland, Belgium and Russia were in attendance to hear a wide range of interesting papers. The forum opened with one-minute of silence for Rex, before new chairman Brian Harvey gave his talk on little-known Soviet satellites exploring more exotic areas of science such as earthquake prediction and the secrets of "anti-matter".

Although the US Space Shuttle has just delivered a large anti-matter detector to the ISS, Brian reminded us that similar instruments had been flown by the Soviets as far back as the 1960s but – unfortunately - this is often forgotten because the science came from the "wrong side" of the Iron Curtain. Whilst researching a recent book on the subject, he was surprised to discover that data collected over 40 years ago by the Proton heavy cosmic ray satellites is still being used today.

Bert Vis gave two separate photographic slideshows of a recent visit to the cosmonaut training centre in Moscow. Luckily personal contacts there ensured that he had a rare invitation to the official 50th anniversary Yuri Gagarin celebrations in April and he was pleasantly surprised to report that previously neglected buildings and space monuments there had been renovated for the occasion.

Although Prime Minister Putin was rumoured to be attending the Star City event, Bert told us he knew it wasn't true by the relative lack of "heavies" around. He also wryly observed that the growing presence of new CCTV cameras was making him more nervous about trying to photograph more obscure corners of the base!

Next up was Bart Hendrickx, who gave a detailed account of future Russian plans for manned spaceflight. Using official graphs and diagrams, he told the audience about: new Russian ISS modules; a post-Soyuz six-seat reusable capsule (its designers had originally wanted to land using only retro-rockets for landing but the cosmonauts insisted on parachutes); a new launch site to be located in eastern Russia; and a proposed $150m per seat space tourist flight to the Moon.

Orbital 'situational awareness' expert Fritz Muse then updated the story of global attempts to track the increasing number of objects in Earth orbit. Although he said ideally satellites should only be placed in space for 25 years, there was now a growing realisation that a lot of debris was destined to remain for thousands of years.

After a fine lunch, Commercial Space Technology's Gerry Webb spoke about the increasingly commercially-minded Russian rocket industry. CST brokered the launch of its first piggyback payload on a Russian rocket in 1995 and has so far placed 24 satellites on five different types of booster. Although the Strela, Rokot, and Dnepr boosters are to be retired by 2017, he believes the enthusiasm and technical skill still exists to develop a new range of small satellite launchers in the very near future.

The last talk of the day was by Pat Norris and focussed on Chinese attempts to leapfrog a technical gap when it comes to its spy satellites. China is increasing its military budget to superpower levels and he noted that they have quickly managed to up-grade from old "wet film" spy technology to more advanced electronic optical systems. Although China orbited only one spy satellite per year during the 1980 and 90s, it has so far launched eleven of its newer generation polar orbiting radar/imaging spy satellites in the last five years.

After this final presentation, the forum audience was asked to re-arrange their seats into a large circle for a more informal discussion about the life and legacy of the late Rex Hall. Numerous intimate anecdotes about first meetings and/or shared space sleuthing revealed new aspects to a man who it can be said inspired most of those present in the room in their own interest in Soviet space history. Those present also took a moment to remember the late Neville Kidger and Dutch space sleuth Chris van den Berg, who had passed away only the previous weekend. As we toasted Rex's memory with a shot of Vodka, he would have been proud to know that his beloved Soviet Forum was once again being enjoyed by all those in attendance.

******

Russia still aims for Mars

It is only when you notice the bronze bust of Russian space pioneer Konstantin Tsiolkovsky staring at you from the corner of the reception hall at the Institute of Bio-Medical Problems (IBMP) that you realise its staff are aiming high. Although this large anonymous building situated in the suburbs of Moscow may literally be a world away from Mars, when cosmonauts finally reach its surface they will have done so because of decades of research carried out there.

Ever since its first cosmonauts were selected in 1960, Russia invested heavily in space medicine, with its team of 'space doctors' making sure that not only are the right people chosen but that they remain healthy before, during and after their spaceflight. This responsibility includes everything from the initial selection process, training, in-flight monitoring of cosmonauts, to helping them through life-threatening emergency situations.

For the last 25 years the Institute has been run by its energetic director Anatoly Grigoriev, a man now in his mid-60s but who still manages to impress visitors with his energy and passion. This probably stems from his own introduction to space medicine as he was personally hired by Boris Yegorov, the world's first doctor in space when he flew as part of the Voskhod 2 crew in 1964. Grigoriev himself had to undergo the same medical examinations as the cosmonauts to get his job but any hopes that he might be up-graded were dashed when the official 'doctor cosmonaut' team was disbanded shortly afterwards.

Although IBMP staff have over the years been given a virtual veto on who has the 'right stuff', their own opportunities to fly have been small - so far only five dedicated Russian doctors have made it into orbit. The IBMP's staff are mainly Earth-based, relying on medical tests carried out on cosmonauts in orbit and test subjects on the ground.

Anatoly Grigoriev in 2003

The core of this data is based on the results of experiments carried out aboard the Mir space station in the 1980s and 1990s. In total over 1,500 dedicated medical experiments were carried out by the cosmonauts aboard Mir during nearly 5,000-days of permanent occupation. During that period (seen by many as the golden era of the Soviet space programme) the IBMP team grew to 700 staff in 17 departments.

Now that present crews aboard the troubled International Space Station (ISS) are limited to six-month flights, due to its delayed construction period because of limited Shuttle flights, these long-duration spaceflight experiments might not be matched for another decade or more.

Using this database Russia has developed a winning system for healthy spaceflight that includes; varied and interesting diets; the Chibris gravity-simulation suit, which created a vacuum around the wearer's legs to literally suck pooled blood away from their upper-body as gravity does back on Earth; an elasticated flight suit which give the cosmonaut a 'work-out' as they moved around the space station; and conventional equipment such as a running treadmill and an exercise bicycle. Special medicines have also been developed to be taken before, during and after flights to help cosmonauts adapt to weightlessness and then back again to gravity when they return.

The ultimate aim is to make sure the human body keeps-up with the technology, so that it is medically possible for people to safely undertake a long-duration trip to Mars when such a mission is finally given the go-ahead.

Indeed one of that IBMP team is uniquely qualified to study this data. Doctor Valeriy Polyakov holds the record for the longest duration space mission ever – having spent 437 continuous days aboard Mir in the mid-1990s. At the end of his flight he even made a conscious effort to walk away from his capsule to prove cosmonauts who land on the surface of Mars will be able to do it.

Although Polyakov is now too old to fly in space again there are also other IBMP staff 'on call' who have undergone cosmonaut training and can be included as part of a crew. The latest of these is 34-year-old Sergey Ryazansky, who passed the cosmonaut training course at Star City and is now an official candidate for a future mission to the ISS.

As well as being one of the most experienced of the younger IBMP researchers his selection is fitting as his grandfather was a space designer and one of the original ten signatories of the petition asking the Soviet government for permission to launch Yuri Gagarin on his historic mission.

"I wanted to join the cosmonaut team because of a family tradition – my grandfather was a rocket engineer and we were proud of that connection. It was a dream of mine to work with cosmonauts and I joined IBMP after graduating from the biological department of the Moscow State University," says Ryazansky.

Once construction of the ISS is finally complete and its crew numbers increase, Ryazansky would be ready to fly as soon as he was given the go-ahead. To complement its in-orbit data the IBMP has also developed many innovative ways to simulate space conditions (both physical and psychological) on the ground. The facility boasts a 9-meter centrifuge, pressure chambers and an invention of its own design which can mimic the effects of weightlessness right here on the Earth. Previously test subjects had been immersed in water baths to fool their bodies into thinking they were floating (as any unsupported state creates similar effects to weightlessness in the body) but they often complained of the harsh reactions of their skin to the water.

Although the longest of these 'wet tests' lasted an amazing 56-days, eventually a new 'dry immersion tank' was developed. This comprises of a water-tight plastic membrane layer situated between the tester and the water, allowing them to 'float' on liquid but still remain dry - keeping skin problems and complaints to a minimum.

For more dangerous experiments space doctors of all nations often use animals. At the start of their space programme the Russians were fond of dogs (their mental reactions are similar) but they decided to follow the American practice of using primates in the early 1980s because a monkey's physical reaction to spaceflight is nearly identical to a human. The first such mission carried a monkey called 'Bion' on Cosmos 1514 in 1983. The series has been called Bion ever since and since then over a dozen multi-national bio-medical missions containing primates, rats and newts have flown successfully.

Importantly for humans these animal flights have proved conclusively that living organisms exposed to long-term microgravity won't suffer any type of permanent cell damage, hereditary gene mutations or chronic stress. Any changes in muscles, bones, cardiovascular and any neuro-sensory systems are only short-term and reversible.

Originally constructed in 1970s the centre-piece of the IBMP facility is a large multi-module space station mock-up we now know is based on a secret effort to design a manned Mars vehicle in the 1960s called the TMK. This project had hoped to stage a manned fly-by of Mars by cosmonauts (but not a landing) in the 1970s. Incredible as it may now seem, if resources hadn't been diverted away from the project to try catch Apollo challenge the Russians had hoped to attempt this mission by 1975!

After they spectacularly lost the 'Moon Race' the Russians reverted back to a more conservative approach, slowly developing new technology and more detailed medical data aboard a series of space stations. Unfortunately the dire economic situation caused by the collapse of the Soviet Union ruled another shot at a manned Mars mission out once again.

Consequently over the years this impressive simulator has been the site of over half-a-dozen long-duration Earth-based simulations. The longest so far being a 240-day isolation test in 1999, whilst the latest is called 'Mars-500.'

'Mars-500' isolation test module

During this new experiment an international team of six researchers is being isolated inside for a total of 520-days - representing a 250-day 'journey' to Mars, a 30-day trip to the surface (during which time the researchers will live on a 'Martian' surface) and a final 240-day 'return' to Earth.

To add to the realism all communication with the outside will be via email, as on a real Mars mission any meaningful voice communication would be impossible because of the time lag for radio messages to get from the ship to the Earth and back.

Russian space doctors are also well-versed in the study of human psychology aboard space stations – something that is often more difficult to predict than a person's physical reactions. Bizarrely, ground-based experiments have often provided the most interesting data. In the mid-1980s scientists started to simulate the effect of upper-body blood pooling (a problem for space travelers as their hearts don't need to fight against gravity to circulate their blood to the 'lower' parts of their body) by tilting Earth-bound testers at six degrees to the horizontal.

During the longest test, a mammoth 370-day 'bed rest' study that began in 1986, things began to go wrong from an early stage. Researchers soon began to realise that the interactions between the experimental subjects were their main worry. Although the test subjects had been grouped into 'crews' of five, three and two in separate rooms they all started to argue amongst themselves. It got so bad amongst the large five-man group that one of the most vocal was "expelled" to one of the smaller groups. The experiment had long-term consequences too. One of the test subjects is reported to have fallen in love with the female nurse who looked after him during the year-long experiment and several others divorced their wives when it was all over.

Another group interaction experiment at IBMP that didn't go according to plan was the first international "SFINCSS" isolation test carried out at IBMP during 1999-2000. During that 240-day experiment a Canadian female participant accused a Russian male colleague of sexual harassment, whilst a Japanese researcher left early over issues of privacy. Although embarrassed about the circumstances Russian researchers believe these situations only highlight the fact that cultural traits might be a larger problem to overcome on long-duration international missions than the medical issues.

So if an isolated crew inevitably turns on itself, would a mixed of male-female crew make the situation better or worse? Dr. Valeriy Polyakov has made his own views known and was even quoted as saying that women should be excluded from the first missions because they are an 'unnecessary emotional and hormonal disturbance.'

Women cosmonauts have been a rarity in the past - perhaps reflecting a long-standing Russian attitude to their place in society - but unfortunately this has resulted in so few of them having flown long-duration missions that the physical effects on the female body in orbit is less well understood. "There are a lot of health problems for women in space," admits Anatoly Grigoriev. "Of course some that are different from men, because women have different metabolisms and hormones but that are enough for them to have serious problems adapting to the space environment."

The only mixed-sex international mission in orbit that seems to have worked well was astronaut Shannon Lucid's 188-days stay aboard Mir in 1996. But this only seems to have worked because of a maternal relationship that grew between herself and her younger crewmates.

"I should say she adapted very well - there were no problems psychologically or with her health. She called the two men in her team her 'sons' and they called her 'grandma'," revealed Grigoriev. Years of research now shows that good morale is the most effective weapon against problems in orbit.

To keep in constant contact with cosmonauts aboard the ISS there is a small 'mini-mission control' at IBMP from which staff use this to monitor the comfort of the crew. They also encourage relatives of the cosmonauts to visit and talk with their loved ones.

"The secret to keeping people healthy in orbit is threefold: preventing their muscles from wasting away because of a lack of gravity to work against; monitoring the heart and making sure it stays healthy; and making sure the crew's mental health doesn't suffer because of the isolation," says director Grigoriev. In so-called "active zone" situations (i.e. during launch or a spacewalk) a cosmonaut's health is monitored in real-time, with continuous measurements of heart rate, blood pressure and breathing, and a routine medical examination every ten days. As one of the doctors in the control centre quipped to me: "They can't imply that they lack attention!"

A full medical kit is carried aboard the ISS for all possible emergency situations but the IBMP doctors have the authority to end a flight if they think there is a medical emergency in progress. This option has only been taken once in the past, when cosmonaut Vladimir Vasyutin was evacuated from the Salyut 7 space station after he suffered a potentially life-threatening infection. Grigoriev reminds us that on a Mars mission the crew will have to fend for themselves if a serious medical problem arises: "The flight is going to be extremely long. On a flight around the Earth if there is something wrong with the cosmonaut we have an opportunity to bring him home. This situation is completely different on a flight to Mars."

For this reason there will almost certainly be a doctor on the first Mars crew. Another major worry that could endanger the crew will be cosmic radiation - something which we will have to try and protect the crew against. This is very important because they are leaving the Earth's own magnetosphere which protects us from the Sun's rays and other dangerous particles.

Although much can be done to physically protect the crew from radiation - such as surrounding the living quarters of a Mars-ship with the fuel tanks and other items to try and absorb some of the radiation - a more exotic solution might be the development of an artificial electromagnetic device that would create a 'force field' around the ship.

Probably the most effective tool will be a network of monitoring stations on Earth to detect any violent outbursts from the Sun and give the Mars crew a few minutes warning so they can retreat to a safe zone onboard as the radiation particles reach them.

IBMP researchers seem to be of the view that a crew can't be protected 100 percent and thus believe the first cosmonauts to Mars should be mature people in their 50s, who have already had families and don't need to worry about any possible long-term genetic damage they might be exposed to.

Russia (or any other nation for that matter) probably couldn't afford a solo manned mission to Mars now or in the immediate future. If it is to happen sooner rather than later, the Russians seem willing to take part in an international expedition. Despite the changing political climate, the relationship between the space medicine teams of the East and West has always been good - something which is obvious everyday aboard the ISS.

"We don't just have a professional interest in each other, there is a personal trust," reminds Grigoriev. "Perhaps that is why we have been co-operating in space medicine since 1971, even through the so-called 'Cold War' period. The only group that seemed to tie our two countries together during that time was the one working on biomedical space problems."

"Of course we have different education systems and different cultures. In spite of the fact that we have some differences in our medical approaches both sides manage to use all that we know to make our work better. That is why a new system was born that is a mix of the two systems. An example of this is the use of the treadmill – we believe that the main way to prevent problems though the flight is to use this, while our American colleagues often believe that the most important exercises in space are 'resistive exercises'. So onboard the ISS we have both measures, giving the cosmonaut the choice."

The institute is also beginning to forge links with China, the newest manned space power, and has co-operated on some satellite experiments. Future cosmonaut Sergei Ryazansky is all in favour of bringing the strengths of each space nation together: "I think it must be a multi-national mission because it will be easier politically, financially and technologically. It is well-known that in some fields Russians are better, in some fields US engineering is better, and now in computer support Japanese companies have much better technology. The best thing would be to combine them."

After they lost the Moon Race, Russian eyes focused back on Mars before their recent dire economic situation seemed to make that an impossible dream once again. But now with a revival in fortunes for the Russia space programme, brought about by its innovative embrace of commercialism using tried and tested space technology, a Mars mission might be on the cards again.

It was formally adopted into the Federal Space programme when a report called 'Manned Mission to Mars' was commissioned. Although only a paper study looking at the technology issues such as the 'liquid fuels versus nuclear engine' debate its existence has signaled a new seriousness towards the topic.

"I think it will be in the third decade of this century. I intend to see it!" proclaims Grigoriev confidently. He is firm in his belief that the only realistic option will be an international mission. When describing such a flight cosmonaut candidate Sergei Ryazansky says it all: "I just want to see this flight, not just to fly personally, but to finally see that human beings are able to go to Mars!"

One can't but be impressed by the passion and commitment of the IBMP staff. Many of them have dedicated their lives to the space programme - seeing both its high and low points - but they now sincerely believe that we are once again picking-up the momentum for a manned mission to Mars sooner rather than later.

******

My IBMP interviews were conducted in 2003. Grigoriev retired shortly afterwards, whilst Ryazansky joined a 105-day isolation 'test crew' before flying to the International Space Station twice. One day he might even fulfill his dream and get to Mars!

Copyright

About the Author

Dominic Phelan was born in Dublin in 1972. He is the editor of *Cold War Space Sleuths* (Springer-Praxis 2013) and the author of *Cornelius Ryan: D-Day War Reporter* (2014).

www.ingramcontent.com/pod-product-compliance
Lightning Source LLC
Chambersburg PA
CBHW021005180526
45163CB00005B/1902